Oceanographic Analysis with R

Dan E. Kelley

Oceanographic Analysis with R

 Springer

Dan E. Kelley
Oceanography
Dalhousie University
Halifax, NS, Canada

ISBN 978-1-4939-9401-4   ISBN 978-1-4939-8844-0   (eBook)
https://doi.org/10.1007/978-1-4939-8844-0

This Springer imprint is published by the registered company Springer Science+Business Media, LLC, part of Springer Nature.
The registered company address is: 233 Spring Street, New York, NY 10013, U.S.A.

For Mum and Dad.

Preface

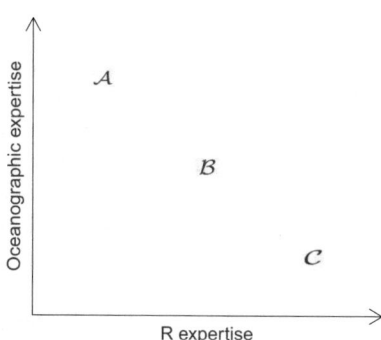

I wrote this book with three types of reader in mind. Type A consists mainly of experienced oceanographers[1] who want to add R to their list of tools for data analysis. They should expect to learn how R can simplify and energize their research programmes. In type B are technicians, consultants and others who may be called upon to solve new problems without sufficient time to study the literature or to implement clean algorithms. Since R provides a wide range of well-vetted solutions that are tied closely to the literature, readers of type B (and their employers) can look forward to an increase in productivity and a reduction in stress. Finally, type C comprises students who are entering oceanography, equipped with R skills gained during previous studies. I hope they will see how to use their skills to competitive advantage during the transition to oceanography.

Taken together, these three types make up the core of any department of oceanography, spanning ages, skills and interests. Given their varied backgrounds and ambitions, they may have different reasons to read this book, and so it is organized with this in mind.

---

[1] Limnologists should also find the book useful, since many instruments and methods are shared between the fields. However, it is tiresome to read "oceanography and limnology" repeatedly, and OAR is a pleasingly nautical abbreviation for the book title.

Chapter 1 puts R in the context of other tools used in oceanography, as part of an argument that now is a good time for oceanographers to try R. Although this chapter is intended mainly for readers of types A and B, who tend to be accustomed to old tools and wary of new ones, readers of type C may find it useful for setting the scene.

Chapter 2 provides an R tutorial that is framed in oceanographic examples. This should be examined closely by readers of types A and B, who will need to learn R syntax. Readers of type C might choose to skim this chapter, with an eye to the peculiarities of oceanographic data.

Chapter 3 presents a sketch of the oce package. Since it was developed during the evolution of a research programme, oce is a very practical thing. It provides functions for a wide range of oceanographic computations, for reading dozens of instrument-specific oceanographic data formats, and for producing graphics that obey oceanographic conventions (Kelley and Richards 2018). Its object orientation scheme lets analysts work at a high level of abstraction, without losing the ability to probe lower levels when appropriate. Although there are excellent tools for individual tasks in other computing languages, few match oce as a coherent framework. Also, and very importantly, the decision to use R means that users have access to thousands of packages for statistical and other operations, yielding cutting-edge methodologies without the burden of extensive coding.

Chapter 4 contains explanations of how R might be used in real-world applications. Here, the steps of the analysis are explained in detail, from start to finish. Drawn from the classic oceanographic literature, the applications sample the four sub-disciplines of oceanography: chemical, biological, geological and physical. While practical-minded readers might focus on the R code, I hope that any reader with an interest in oceanography will welcome the chance to explore data put forward by the likes of Alfred Redfield, Gordon Riley, Tuzo Wilson and Walter Munk.

Chapter 5 continues the applied theme, but with less depth and more breadth. A miscellany of methods, this chapter is likely to be consulted a section at a time, as needs arise. Readers of type C should note that oceanography has yet to develop standard operating procedures, and so this chapter is more a suggestive guidebook than a detailed map.

Chapter 6 provides solutions to the many exercises that pepper the text. This is a key element of the book, because working on exercises is a sure way to build skill. Little is gained by passive reading ... nobody ever learned to play a violin by watching someone else play one.

Appendix A contains advice to readers who are switching to R from Matlab. Appendix B has an outline of popular GUI systems that simplify the use of R, without limiting its power. Appendix C holds a discussion of map projections in the oce package, while Appendix D explains how oce lets analysts switch between the UNESCO and GSW formulations of seawater properties. A few aspects of high-performance calculations in R are sketched in Appendix E. Finally, in Appendix F, readers will find some remarks on the future of R, in general terms and in the context of oceanographic analysis.

Colophon. This book was typeset in LaTeX, with R being used for all the diagrams. The code for the diagrams, along with all the sample code, is embedded in the text using Sweave (Leisch 2002), with R acting as a preprocessor that creates output and diagrams prior to typesetting. This setup means that readers can be confident that R code provided in this book will work as indicated.

Acknowledgments

Many people have helped me with R, with the development of the oce library, and with the writing of this book. It gives me great pleasure to thank, in alphabetical order: Jakson Alves de Aquino, for writing a powerful and graceful R/Sweave plugin for the Vim text editor; Paul Barker, for helping with GSW test cases; Natacha Bernier, for leading field trips in my physical oceanography class at Dalhousie University, including one that yielded a dataset in oce; Daniel Bourgault, for long-time collaborations; Richard Cheel, for providing advice on dealing with acoustic instruments; Alex Deckmyn, for helping with wind roses in oce; Pierre Flament, for contributing his seawater spiciness code to oce; Peter Galbraith, for being my first PhD student, and a steadfast collaborator ever since; Jenna Hare, for help with French oce plot labels; Alex Hay, for collaborating in SLEIWEX and beyond; Jackie Hurst, for helping with typesetting and dealing with university administration; Stephanie Kienast, for useful R discussions and for help with German oce plot labels; Anthony Ladson, for helping with tidal analysis in oce; Paul and Sue Logan, for explaining the Rink Ratz® hockey card game; Chantelle Layton, for contributing to oce and for spearheading research relating to Flemish Cap and baroclinic topographic Rossby Waves; Chris Moore, my Dean at Dalhousie University, for suggesting it was reasonable to take the time to write a book; Edward P. Morris, for helping to make oce handle burst-mode velocimeter data; Michelle Paon, for helping me to assemble the citation data for Chap. 4; Eric Mills, for providing insights on Gordon Riley and for sharing material from his excellent history of the development of oceanographic science (Mills 2009); Ramzi Mirshak, for SLEIWEX collaborations, including deep thinking on beamwise ADP calculations, which benefited oce after the fact; Rich Pawlowicz, for sharing Matlab solutions to a seemingly unending series of oceanographic problems; Clark Richards, for his SLEIWEX collaborations, for his coauthoring of both oce and gsw, for his insightful comments on this book and for his wide-ranging dedication to science and education; Doug Schillinger, for patient explanations of the functioning of acoustics-based instruments; Christopher Taggart, for his open door and tireless intellectual spirit; and Keith Thompson, for advice on statistics, science and more important things.

More generally, I am indebted to Dirk Eddelbuettel, Uwe Ligges, Brian Ripley and Hadley Wickham for their remarkably generous contributions to R community, and the many others who have written and maintained the R packages on which users rely.

Finally, and most importantly, I extend the warmest of thanks to the many students, undergraduate and postgraduate alike, with whom I have had engaging conversations about the place of R in oceanography. I've learned a lot from them, and I hope to return the favour in the pages that follow.

Halifax, NS, Canada   Dan E. Kelley
Spring 2018

Contents

Acronyms

ADCP	Acoustic Doppler Current Profiler. This acronym was coined by RDI-Teledyne; similar instruments made by Sontek are called acoustic Doppler profilers, with the acronym ADP. In the oce package, the adp class holds both types of data
ADV	Acoustic Doppler Velocimeter, a device that uses acoustic signals to infer velocity at a point. This acronym is used by Sontek; Nortek refers to similar instruments as "Vector velocimeters." In oce, the adv class holds both types
CRAN	Comprehensive R Archive Network, the repository for the R application and associated packages[1]
CSV	Comma-Separated Value, as used in spreadsheets
CTD	Conductivity-Temperature-Depth instrument. In the oce package, data from CTD instruments are of class ctd
EOF	Empirical Orthogonal Function
GEOSECS	Geochemical Ocean Section Study
GSW	Gibbs SeaWater toolbox of TEOS-10, provided with the gsw package and various oce functions; see Sect. 5.2.1 and Appendix D
LISST	Laser In Situ Scattering and Transmissometry instrument
MEDS	Marine Environmental Data Service[2]
NetCDF	A self-describing binary data format used in some applications in oceanography and atmospheric science[3]
NOAA	US National Oceanographic and Atmospheric Administration[4]
NODC	US National Oceanographic Data Center[5]

---

[1] http://cran.r-project.org.

[2] http://www.meds-sdmm.dfo-mpo.gc.ca/isdm-gdsi/index-eng.html.

[3] http://www.unidata.ucar.edu/software/netcdf/docs/.

[4] http://www.noaa.gov/.

[5] http://www.nodc.noaa.gov/.

ODF	Ocean Data Format, used by the Canadian Department of Fisheries and Oceans[6]
ODV	Ocean Data Viewer software[7]
POP	Practical (or Provisional) Operating Procedure.
SLEIWEX	St Lawrence Estuary Internal Wave Experiment
SOP	Standard Operating Procedure (contrast with POP).
TEOS-10	Thermodynamic Equation of Seawater-2010[8]
UNESCO	United Nations Educational, Scientific and Cultural Organization
WOCE	World Ocean Circulation Experiment[9]

---

[6] http://slgo.ca/app-sgdo/en/docs_reference/documents.html.

[7] http://odv.awi.de.

[8] http://www.teos-10.org.

[9] http://woce.nodc.noaa.gov/wdiu.

Symbols

The next few pages list some symbols used in oceanography. The R commands used here require the `oce` package to have been loaded, with

```
library(oce)
```

which may be done in a startup file (see page 10). It is also common to use this startup file to specify a default seawater formulation. In this book, most examples use the Gibbs SeaWater (GSW) formulation (McDougall and Barker 2011), as established with

```
options(oceEOS="gsw")
```

in the author's startup file. The older UNESCO system, denoted `"unesco"`, is also available throughout `oce`. See Sect. 5.2.1 and Appendix D for more discussion of these systems, and note that a choice of equation of state can also be made in function calls, as illustrated below.

ρ in situ seawater density in kg/m^3. For example, at practical salinity 35 PSU, in situ temperature $10°C$ and pressure 100 dbar, the UNESCO and TEOS-10 formulations of seawater density are[1]

```
swRho(salinity=35, temperature=10, pressure=100,
     eos="unesco")
```
```
| [1] 1027.404
```
```
swRho(salinity=35, temperature=10, pressure=100,
     longitude=300, latitude=30, eos="gsw")
```
```
| [1] 1027.406
```

(Note that the GSW formulation requires longitude and latitude, and it is a geographical variation of seawater "salt" ion ratios that yields the small density difference seen above.) The names of the arguments could be omitted, e.g.

```
swRho(35, 10, 100, 300, 30, "gsw")
```

works as above. In R, argument names are optional, provided that they are given in the correct order, and without gaps. R also permits abbreviation of argument names, e.g. `t=10` could be written instead of `temperature=10`, as explained in Sect. 2.3.11.2.

---

[1] In a convention employed throughout `oce`, this function starts with "sw" to indicate that it applies to seawater. Analogously, air density may be calculated with `airRho()`.

σ Density anomaly, $\rho - 1000\,\text{kg/m}^3$, calculated with swSigma().

σ_θ Potential density anomaly, referenced to surface pressure;

```
swSigmaTheta(35, 10, 100, eos="unesco")
| [1] 26.95398
```

is equivalent to

```
th <- swTheta(35, 10, 100, eos="unesco")
swSigma(35, th, 0, eos="unesco")
| [1] 26.95398
```

σ_t Crude form of potential density anomaly, defined as $\rho - 1000\,\text{kg/m}^3$, with ρ based on in situ temperature and zero pressure.

```
swSigmaT(35, 10, 100, eos="unesco")
| [1] 26.952
```

$\sigma_0, \ldots, \sigma_4$ Potential density with reference pressure 0 dbar, 1000 dbar, 2000 dbar, 3000 dbar and 4000 dbar.

θ Potential temperature, i.e. the temperature of a water parcel moved adiabatically from one pressure to another, e.g.

```
swTheta(35, 10, 100, eos="unesco")
| [1] 9.988453
```

for movement to the surface, or

```
swTheta(35, 10, 100, 1000, eos="unesco")
| [1] 10.10996
```

for movement to 1000 dbar. (These two calculations illustrate the use of default values for function arguments; see Sect. 2.3.11.2.)

Θ Conservative temperature, as defined in GSW.

CT   Conservative temperature argument name in GSW functions.

f Coriolis parameter, e.g. at 45°N

```
coriolis(45)
| [1] 0.0001031261
```

g Acceleration due to gravity, e.g. at 45°N

```
gravity(45)
| [1] 9.80619
```

N^2 Square of buoyancy frequency defined by $N^2 = -g\,\rho_0^{-1}\partial\rho/\partial z$ where ρ_0 is a reference density. N^2 may be calculated with swN2().

p Sea pressure, i.e. in situ pressure minus atmospheric pressure. Given hydrostatic balance $dp/dz = -\rho g$,

```
gravity() * swRho(35, 10, 1, eos="unesco") / 1e4
| [1] 1.007053
```

illustrates the near equivalence of sea pressure in dbars and depth in m, since 1 dbar is 10^4 Pa.

S Seawater practical salinity, in the UNESCO system.

S_A Seawater absolute salinity, as defined in GSW (Sect. 5.2.1 and Appendix D.)

SA	Absolute salinity argument name in GSW functions.
SP	Practical salinity argument name in GSW functions.
t	Time, for most of the oceanographic literature. (In some thermo-dynamic treatments, e.g. the GSW literature, t stands for in situ temperature, in °C.)
t	In situ temperature argument name in GSW functions.
T	In situ temperature in Celsius, for most of the oceanographic literature, although in Kelvin for some thermodynamic analyses, e.g. in the GSW literature.
u, v, w	Components of velocity in the x, y and z directions.
x, y, z	Horizontal and vertical Cartesian coordinates. Typically z is measured in metres above the mean sea surface. Since instruments measure pressure instead of vertical coordinate, conversion with swZ() or swDepth() can be useful.

Chapter 1
Why R, and Why Now?

Abstract For many years, the R language has had a reputation as a premier system for interactive data analysis. From a user's perspective, there are two main reasons for this. First, R is a language designed specifically for working with data, so it has important practical features (e.g. sensible treatment of missing values) that are not found in more general languages. Second, R comes with a vast array of high-quality packages, or libraries, that handle specialized tasks. The packages are contributed by experts in various fields, and tend to be tied closely to the literature—two facts that are relevant in an integrative field such as oceanography. The case for R has grown stronger in recent years, with a general movement to open-source software, and with specialized aspects of oceanographic data analysis becoming available in the oce package. Now is a good time for oceanographers to try R.

In a young scientific field, work is often carried out by postgraduate students whose thesis goals inspire new procedures intended for somewhat limited application. These procedures might be called practical (or provisional) operating procedures (POP), by analogy to the standardized operating procedures (SOP) used in more routine work. As fields mature, POP may be translated to SOP, expanding the range of application and permitting a shift in workload to technicians who do not need postgraduate training. According to this line of reasoning, new undergraduate programmes can be a sign of a maturing field. This is the state of oceanography today.[1]

The task of translating POP to SOP may be eased if similar tools are used in each, so it makes sense to consider the choice of tools carefully. In this spirit, Fig. 1.1

---

[1] For an example, the author was contributing to the development of a new undergraduate programme at Dalhousie University, while working on this book.

© Springer Science+Business Media, LLC, part of Springer Nature 2018
D. E. Kelley, *Oceanographic Analysis with R*,
https://doi.org/10.1007/978-1-4939-8844-0_1

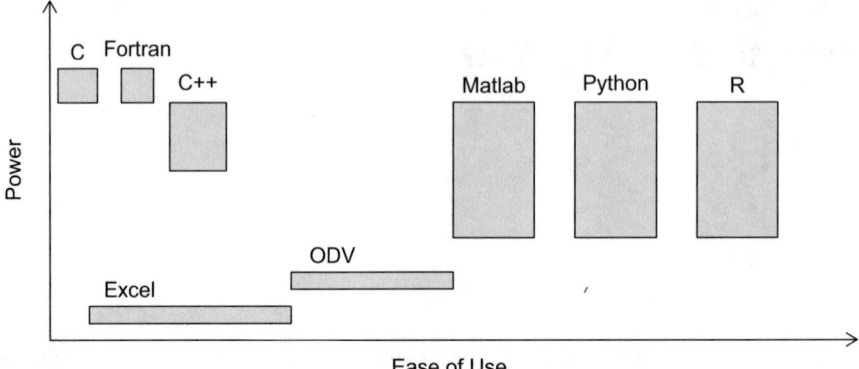

Fig. 1.1 Comparison of general-purpose computing languages or applications that may be used for oceanographic analysis

compares R with some other systems that might be used for oceanographic data analysis.[2] Some of these systems hold little promise, but it is worth touching on them all, if only for the excuse to bring up some general issues.

The diagram suggests that Excel scores poorly on both power and usability. While the first point is unlikely to be contested by anyone who has tried to use Excel on a large dataset, some readers might argue that Excel is easy to use. However, the context is important. Compared with its competitors, Excel is ill-suited to the particular calculations and graphical displays that oceanographers need. For example, it is easy to add columns in Excel, but considerably more difficult to correctly enter a formula for seawater density that contains dozens of numerical values specified to five or more digits. Also, the very thing that makes Excel popular for nontechnical work, its graphical user interface (GUI), is an impediment in technical work,[3] because a sequence of GUI operations is difficult to describe and reproduce.[4] A text-based approach is preferable to a GUI approach for all but the simplest of tasks. Those who switch from Excel to R should see benefits quickly, and should find the transition easy, because there are tools for combining the two systems (Heiberger and Neuwirth 2009).

To some extent, the box for Excel in Fig. 1.1 is a place-holder for other GUI systems, and so these need not be discussed in much detail, with one exception:

---

[2]This book deals more with data analysis than with statistics. For early thoughts on data analysis, see the influential paper by Tukey (1962), along with the recent historical commentary by Mallows (2006).

[3]GUI-based systems can be problematic for users with weak vision, with text-based systems such as R providing a better choice (Godfrey 2013; Godfrey and Erhardt 2014).

[4]Issues in reproducible research are discussed by Pebesma et al. (2012), while Herndon et al. (2013) detail problems particular to Excel.

Ocean Data Viewer.[5] ODV is a GUI-based system that offers good support for many oceanographic operations, including the equation of state, specialized graphing, etc. However, its power is limited by both its GUI-based design and the fact that the ODV source code is not available for inspection or modification.

All the other entries in Fig. 1.1 are languages, some compiled and others interpreted. Languages are well-suited for reproducible research, because the code used to solve a problem is, in and of itself, a full description of the processing procedure. Good coding practices make the transference of effort between tasks or work groups an easy matter, often involving little more than changing the name of a data file. Readers who are accustomed to the GUI approach will discover other benefits in adopting the language approach. Loops make it easy to carry out repetitive work. Conditional blocks handle changing circumstances. Functions and object-orientation yield specialization and simplicity of operation, without loss of generality. The only cost for these benefits is a learning process that starts with thinking beyond menus and icons.

Generally, compiled languages offer higher efficiency than interpreted ones, but they are much more difficult to use. This is why the compiled languages C, Fortran and C++ are placed on the left of Fig. 1.1. These are used in the most demanding of computing tasks, from operating systems to climate models. The relative positions of these languages on the diagram are debatable, since they depend on the nature of the work being carried out. C offers essentially the full power of the machine, but the language is difficult to use for oceanographic work, because of its weak support for matrices and other high-level data types. Fortran offers similar power, and has an advantage over C in its strong support for matrices. In some ways, C++ is even easier to use, with an object orientation model that reduces coding effort and facilitates collaboration, but its object orientation can impose efficiency penalties, if users rely on overly indirect algorithm expression.

Although compiled languages underpin all computing applications, and remain the best solution for large computing tasks such as numerical models, they have fallen out of favour for interactive work. This is particularly true for so-called "exploratory data analysis" as described in the seminal treatment of Tukey (1977) and more recently by, e.g., Velleman and Hoaglin (2004). Of many interpreted languages that might be discussed in the present context, three stand out: Matlab, Python, and R.

As with the compiled languages shown in Fig. 1.1, the relative merits of the interpreted languages depend on the work being done. The illustrated efficiency ranges are large because not all problems map well to the fastest components of the languages. For example, these three languages all provide strong low-level support for matrices, so that problems that can be cast in matrix form are handled with efficiency approaching that of compiled languages. Importantly, each also allows

---

[5]http://data.unep-wcmc.org.

advanced users to frame parts of their algorithms in compiled languages, yielding great improvements in speed.[6]

The most contentious aspect of Fig. 1.1 may be the ranking of Matlab, Python and R in terms of ease of use, for this is the sort of judgement that partly boils down to a matter of taste. The diagram expresses the author's opinion, based on years of experience, that Python is superior to Matlab, and that R is superior to both. This reflects several factors. First, both Python and R are popular in more diverse fields (at least outside oceanography), which means that users of these languages can benefit from the efforts of broad communities of experts. The popularity lies partly in the technical merits of the languages, and partly in their open-source licenses. Especially in a university setting, open-source systems attract talented people who have a habit of sharing their work, and this can lead to nonlinear improvements to the development process.[7] For Python and R, the shared efforts are organized through systems that bundle software code with documentation and test cases. The bundles are called packages in R. These packages are a significant factor in the present judgement of the superiority of R, since they provide the power to tackle a myriad of tasks that come up in oceanographic analysis.

An important package in the oceanographic context is oce. As discussed in Chap. 3 and throughout this book, oce handles dozens of specialized oceanographic data formats, and provides functions for calculations and graphical displays that are specific to oceanography. Its object-oriented approach lets novices get results quickly, without imposing undue limits on experts. Reproducible research is built into the foundation of the package, with a processing log being contained in all oce data objects. Few limitations are imposed on the scope of work done with oce, because the package integrates well with both the base R language and other packages.

Based on factors such as those listed above, the thesis statement of this book is that R is a powerful system for oceanographic analysis, with high potential for open-ended research and more routine technical work. Simply stated, it is a tool that works well, and fits comfortably in the hand.

There *is* a learning process in adopting R, and this book is designed to accelerate that process, in different ways for readers of different backgrounds. The author is a research scientist and an educator, not a salesman, and so the text points out the weaknesses of R, as well as strengths. For many readers, these strengths and weaknesses will be measured against Matlab, and so an early component of the tutorial provided in the next chapter is a brief comparison of the two languages.

---

[6]For example, the oce package (Kelley and Richards 2018) uses C to decode the binary data files produced acoustic Doppler instruments, reducing computation times by orders of magnitude compared with pure R.

[7]See Raymond (2001) for a general discussion of open-source development, Fox (2009) for comments in the R context, and Lowndes et al. (2017) for details of how using R and other open-source tools can enhance reproducibility in ocean science.

Chapter 2
R Tutorial for Oceanographers

Abstract R comes with an excellent tutorial that, like many fine tutorials, tends to be ignored by people with little patience for material presented in a general manner. This is why the present chapter uses oceanographic examples to explain R concepts, and why code makes up so much of the text. The early examples are designed to encourage readers to become comfortable whilst navigating the R documentation, because this skill can be the key to moving from simple examples to real-world applications. The main concepts of R data types and language features are illustrated here in practical terms, with many of the explanations involving graphical representation. Since experienced R users are unlikely to study this chapter in great depth, specialized methods of oceanographic analysis are mainly deferred to succeeding chapters.

2.1 Introduction

R can be deceptive at first, because it handles simple tasks so well that newcomers might wonder if it has the power for advanced work. They need not worry, for R balances simplicity and power in ways both subtle and varied. Some users notice this first in the thoughtful system of default function arguments, through which R achieves simplicity without loss of flexibility. Others will focus on how R uses object orientation methods to generalize tasks, letting users think about science instead of syntax. Those with programming experience will see the benefits of the functional basis of R, and its innovative rules for the scope of variables and the evaluation of expressions. And those working on computationally demanding tasks will appreciate the R interfaces to C, C++ and Fortran, and its handling of multiple-processor systems.

R is a practical language that owes some of its strength to its lineage. Many of its best characteristics can be traced to the S and S-plus languages upon which it was patterned. These earlier systems were well designed at the outset, and were honed by use in advanced research settings (Becker and Chambers 1984; Becker et al. 1988; Chambers and Hastie 1992). R was also born in a research setting, which may

© Springer Science+Business Media, LLC, part of Springer Nature 2018
D. E. Kelley, *Oceanographic Analysis with R*,
https://doi.org/10.1007/978-1-4939-8844-0_2

explain why it has innovations that take it beyond the earlier languages (Ihaka and Gentleman 1996; Chambers 2008).

The important book by Venables and Ripley (1999) contains a wide-ranging and authoritative overview of R and its use, and it can be recommended to any reader. Dalgaard (2002) is a good companion, especially for beginners. The Chambers (2008) and Wickham (2014) treatments of technical aspects should prove useful to advanced users, especially those developing R packages. There are also many books about specialized topics, e.g. graphical display (Murrell 2006; Wickham 2009), time series analysis (Shumway and Stoffer 2006), neural networks (Ripley 1996), Bayesian methods (Albert 2009), numerical ecology (Borcard et al. 2011), etc. Readers should have little difficulty finding books on a specialized applications of R; for example, the present book is part of a Springer "UseR!" series that has dozens of titles.

In addition to texts, the extensive features of R are covered in detail in the official documentation (R Core Team 2017). For beginners, the most important part of this is the essay entitled "An introduction to R." Other essays deal with R as a language, with writing packages to extend the system, etc.; these are recommended to readers who already use R frequently. There is also a full reference manual that, spanning thousands of pages, is best consulted a little at a time.

A sensible way to learn R is to work through a tutorial. The "Introduction to R" can be used in this fashion, and few readers would be disappointed with its pacing, coverage or clarity. The present chapter is not as deep, nor as broad, but it does have two advantages: (a) it is cast in oceanographic terms, which may hold readers' interest better than general material and (b) it contains many exercises that should speed up the learning process.

By the end of this chapter, readers should be able to accomplish simple tasks in R, and understand code for more complicated tasks. This will set the stage for the upcoming chapters, in which the focus shifts more directly to oceanographic analysis. But, before any of this can be done, we must acknowledge the "elephant in the room", Matlab.

Over the past few decades, Matlab has become so popular in oceanography that many regard it as a *lingua franca* for data analysis. Figure 2.1 (and Appendix A) illustrates that mapping from Matlab to R is not difficult, especially when viewed in stages. For example, the first Matlab line

```
load xy.dat
```

causes a file named `xy.dat` to be read, with the numerical values being stored in a matrix named `xy`. Although the syntax is simple, it hides a great deal. Upon encountering the `load` token, Matlab interprets the next token as a file name, and constructs a variable with an analogous name, into which to store the contents of the file. By contrast, the equivalent line in R

```
xy <- read.table("xy.dat")
```

is more complicated, but also more direct. The "`<-`" token indicates assignment. To its left is the name of a variable to store the result of an expression to its right. In this case, the expression is the value returned by a function named `read.table()`

```
load xy.dat
plot(xy(:,1),xy(:,2))
quit
```

```
xy <- read.table("xy.dat")
plot(xy[,1], xy[,2])
q()
```

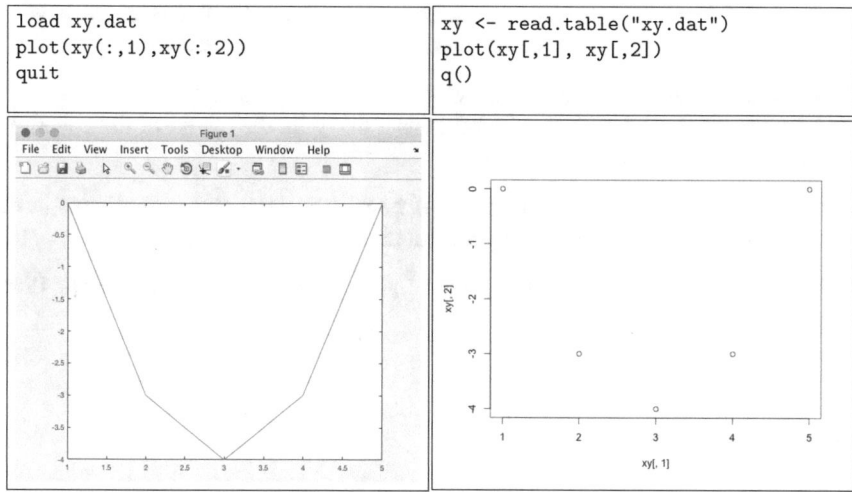

Fig. 2.1 Comparison of Matlab and R for file input and graphics

that reads tabular data and returns a so-called "data frame" (discussed at length later) representing information within the named file.

Next, the Matlab line

```
plot(xy(:,1),xy(:,2))
```

constructs a line graph with the first column of xy taken as the *x* coordinate and the second as the *y* coordinate. The ":" means to use all rows. Note the use of parentheses in two very different ways here, to indicate arguments to a function and to indicate indices of a matrix. The R version

```
plot(xy[,1], xy[,2])
```

differs in several ways. First, it creates a scatter graph by default, although this can be changed to a line graph easily (see Exercise 2.1). Second, the axis labels indicate the names of the plotted items, which is very helpful in exploratory analysis because it reduces the need to change axis titles if variables are changed. Third, R uses square brackets for indexing and does not require the ":" place-holder.

Finally, a user exits Matlab with

```
quit
```

while the equivalent in R is

```
q()
```

which calls a function named q(). Matlab users may find it odd to exit a program by calling a function, but this fits the theme of R, which is a function-oriented language. Indeed, many things that one might think of as commands or operators in Matlab take the form of functions in R, and this has subtle and helpful effects that will become clearer in the remainder of this book.

Exercise 2.1 Type help(plot) in a console, and use the results to see how to draw a line graph instead of a scatter plot. (See page 187 for a solution.)

Exercise 2.2 Consult the documentation for read.table(), to see how to indicate that the first line of the file contains a line with the names of the columns. (See page 187 for a solution.)

Exercise 2.3 Use the mfrow argument of par() to draw multi-panel plots in R, emulating the Matlab subplot command. (See page 188 for a solution.)

Exercise 2.4 Use outer() to emulate the Matlab function meshgrid. (See page 188 for a solution.)

2.2   First Steps with R

2.2.1   License

R is subject to a "GNU General Public License". This has three practical benefits. First, it means that R can be included in linux systems, which ensures distribution within a community of technically minded users who tend to help other users by contributing to online forums and sharing code. Second, R is an open source application, so that users can examine its internal workings, in case they want to evaluate the methods or extend them. And third, R is available free of charge, which has obvious benefits to students, researchers and consultants.

2.2.2   Installation

On linux systems, R is installed in the same way as other software, either using GUI operations or by typing commands in a terminal. On other systems, installation is a simple matter of visiting the R website[1] and installing the appropriate pre-compiled version. Archived versions are also available there.

2.2.3   R Packages

R benefits greatly from a scheme for combining code and documentation into so-called packages that are distributed in the Comprehensive R Archive Network, CRAN, which is available at the R website. As shown in Fig. 2.2, the number of

---

[1] http://cran.r-project.org/.

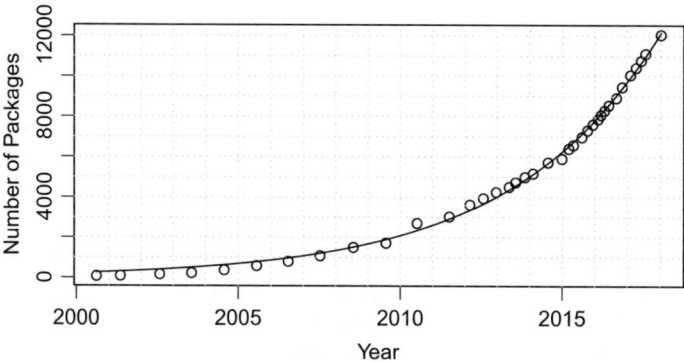

Fig. 2.2 Number of R packages on the CRAN archive site. The curve results from using nls() to perform a nonlinear regression with model $n = n_0 \exp((t - t_0)/\tau)$, where n is package count, $t - t_0$ is the time since the first point shown, and n_0 and τ are free parameters. This yields a doubling time of 3.2 ± 0.1 years (95% confidence interval)

packages[2] has increased dramatically since R was released, with over ten thousand being available at the time of writing. The coverage of these packages is broad, reflecting the broad popularity of R. Since packages are usually developed by experts in the sub-field of application, they tend to address relevant problems in up-to-date ways. Code quality tends to be high, given both the expertise of the developers and automated code-quality checks that are built into the packaging process.

The R documentation explains the process of package development in clear terms, and so it is not difficult for users to develop packages for their own work. Since all R users benefit from the work of others, there is a natural tendency to share. This sharing often starts within work groups, but as code becomes robust, it can make sense to share more broadly on CRAN.

Some of packages used in this book are unlikely to be installed by default in the version of R on the reader's computer. To see whether a package is already installed, open an R console and type, e.g.

```
library(oce)
```

If an error message results, then it will be necessary to install oce, either with a menu item in a GUI version of R or by writing

```
install.packages("oce")
```

in an R console. Development versions of packages may be installed from source files.

---

[2]The package count was inferred from web archives at http://wayback.archive.org/.

For packages that are developed on the GitHub social coding website[3], the building process is, e.g.

```
library(devtools)
install_github("oce", "dankelley", "development")
```

It is important to note that many of the examples in this book deal with the oce package, but the

```
library(oce)
```

that should precede the use of oce has been removed from the code examples in the text, some of which are only a few lines long. However, a library() call will be provided for all other packages used in examples.

Code presented in this book uses packages named boot, bvpSolve, changepoint, DBI, deSolve, devtools, doMC, doParallel, fields, geonames, gsl, hexbin, imputeTS, jpeg, KernSmooth, lattice, lmodel2, lubridate, magrittr, MASS, microbenchmark, mixtools, ncdf4, oce, ocedata, party, plyr, propagate, R.matlab, ReacTran, rgdal, rootSolve, RSQLite, segmented, signal, smatr, tiff, vioplot, WaveletComp, and XML, plus the packages upon which these depend.

2.2.4   Running R

Starting an R console is system-dependent. Many systems provide an icon that can be clicked to launch a GUI-based interface to R (see Appendix B), in addition to a command-line tool that can be used within a terminal.

The startup behaviour of R can be customized with a startup file, e.g. the author has

```
options(digits=7, digits.secs=3)
options(editor="mvim")
options(oceEOS="gsw")
```

in a file called .Rprofile in his home directory, to control the number of digits in numbers and times, to set his preferred text editor, and to default to the "Gibbs Seawater" formulation of the seawater equation of state, as opposed to the older UNESCO formulation (see Sect. 5.2.1 and Appendix D). Information on R startup is provided by help(Startup).

There are GUI wrappers that simplify some R operations, including the popular Rstudio system, but R is still a computing language at its core, with detailed actions being controlled by textual instructions. These instructions may be typed in an R console or entered in a file that is processed by R. Many users combine the two methods, using the console to test provisional approaches, and gradually copying working code into an editor window. Power users are inclined to prefer working in

---

[3] github.com.

editors that are external to GUI applications, and this is easy with `Rstudio`. The most popular standalone text editors are Emacs (for which the ess mode handles R) and Vim (for which the `vim-r` plugin handles R); each allows transferral of individual lines or blocks of code to a running R session.

Except for trivial work, it is a mistake to use R in a solely interactive mode. Saving code in files is necessary to achieve reproducible results. These files should be self-contained, so that they can be used directly by another researcher with access to the data. There are several ways to use such a file. Within an R session, such a file (called `work.R`, say) can be executed with

```
source("work.R")
```

in an R console, by clicking an icon in an `Rstudio` window, or with

```
R --no-save < work.R
```

or

```
Rscript work.R
```

within a unix-like operating system.

It is easy to automate R analysis considerably on Unix. For example, if the following[4] is put into a file called `Makefile`, then typing make in a terminal will run R on all the files with name ending in `.R`, generating a `.out` for each.

```
R = $(wildcard *.R)
OUT = $(R:.R=.out)
%.out: %.R
        R --no-save < $< > $@
all: $(OUT)
clean:
        rm -f *.out
```

Typing make again does not call R, because the `.out` files are older than their `.R` sources. If any of the `.R` files are changed, then typing make again processes just those files. Typing make clean removes the `.out` files, resetting the process. Using make speeds up complex multi-staged work, with the extra benefit (compared with interactive analysis) of documenting the entire processing procedure, a key component of reproducible research.

2.2.5 Getting Help

Readers who have followed along with the worked exercises will already be comfortable accessing the R help system in a basic way, e.g.

```
help(read.table)
```

explains `read.table()`. As a convenience, this can also be written

```
?read.table
```

---

[4]The indented lines of the `Makefile` must start with a tab character, not spaces.

If the function (or dataset) name is not known, it may still be found by searching the documentation, e.g.

```
help.search("trig")
```

yields information on a variety of functions relating to trigonometry and

```
help.search("angle", package="oce")
```

reveals `oce` functions or datasets whose documentation contains the word `angle`. The search string must be enclosed in quotes for `help.search()`, but quotes are optional for `help()`.

An overview of the functions in a package is easily obtained, e.g.

```
help(package="oce")
```

yields the documentation at a broad level, and

```
package?oce
```

yields a more specific entry about the package.

The `example()` function is a companion to `help()` that runs any examples that are provided in the documentation for a named function, e.g. `example(plot)` provides a few examples of plotting. This can be very useful, because many documentation pages have pertinent examples.

Exercise 2.5 Use `help.find()` to find an R package that accesses the www. geonames.org website, and thus locate Halifax, Nova Scotia. (See page 189 for a solution.)

2.3   Syntax

2.3.1   Expressions

R can be used as a calculator, simply by typing expressions, and e.g.

```
377 / 120
[1] 3.141667
```

(Ptolemy's approximation of π) demonstrates a call-and-response typographic convention used throughout this book. Importantly, the response displayed here was created by R itself, using a system called "sweave" (Leisch 2002). Thus, readers can rest assured that the numerical and graphical examples of the book will work as indicated, apart from formatting nuances.[5]

A confusing aspect is the string "`[1]`" preceding the result of the calculation. This is a counter indicating that the first number on the line is the first in a sequence. These counters are helpful for long sequences. For example, the colon operator (`:`) forms a sequence of values (a "vector", in R parlance) in a stated range, e.g. `1:7`

---

[5]R displays a prompt before the input, but this is omitted throughout this book.

yields the integers from 1 to 7 and their inverse cubes can be calculated with the
power operator (^)

```
1 / (1:7)^3
[1] 1.000000000 0.125000000 0.037037037 0.015625000
[5] 0.008000000 0.004629630 0.002915452
```

Even this short vector may reveal the usefulness of output counters.

Readers who skipped the parentheses in the previous expression will see a display
of hundreds of numbers. This is because the exponential operator is acted upon
before the sequence-forming operator. It is said that "^" takes precedence over
":". By contrast, ":" takes precedence over multiplication, as the reader can verify
by entering "1 / 1:7 * 3" in a console. The precedence of operators is well
documented, but it is also easy to forget, so parentheses are recommended to avoid
confusion.

On a more aesthetic note, it is good idea to use white space to clarify the notation,
especially in complicated expressions. This is somewhat a matter of preference, with
some R users inserting a single space before and after every operator, and others
inserting space only for certain operators, or in certain circumstances, depending on
whether the expression is being entered interactively or being inserted in a document
to be shared widely.

Much more could be said about the syntax of R, but it is preferable to continue
on with practical examples. For any but the simplest of work, it is certain that users
will need to make use of R functions. Of course, given its focus on data analysis,
R provides many functions for statistical and numerical work. Most functions can
work with single values, or collections of values such as vectors. This latter fact
will be a commonplace for Matlab programmers, but those coming to R from C-like
languages will find that it greatly simplifies coding, eliminating loops. For example,
the Taylor series approximation

$$e^x = \sum_{n=0}^{\infty} \frac{x^n}{n!} \tag{2.1}$$

for $x = 1/10$ can be evaluated to five terms with

```
sum(0.1^(0:4) / factorial(0:4))
[1] 1.105171
```

where factorial() calculates $n!$ and sum() adds elements.

In addition to such numerical-analysis functions, a conventional set of functions
for programming is available. For example, the minimum of a vector is calculated
with min(), while the index of the smallest element is given by which.min().
The corresponding functions for maximum are max() and which.max(). The
range of values in a vector (or any other collection of numerical values) is calculated
with range(). The boolean function all() examines a vector of boolean values
(or expressions) and returns TRUE if each of these values is TRUE. Similarly,
any() indicates whether any of the values is TRUE. All of these functions can
handle missing values (see Sect. 2.3.3.6).

Comments may be added to R expressions by placing a # character on the line, so long as that character is not in a quoted character string. Since an empty line is a valid R expression, this also provides a scheme for standalone comments.

```
# Note that atan() returns an angle in radians
4 * atan(1)                                    # pi
| [1] 3.141593
```

Exercise 2.6 Use cumsum() to monitor the convergence of the Taylor series for exp(). (See page 189 for a solution.)

2.3.2 Variables

2.3.2.1 Variable Assignment

As noted previously, R denotes assignment with the operator "<-", e.g.

```
phi <- (1 + sqrt(5)) / 2
```

stores the golden ratio in a variable named phi. It is also possible to assign to variables indirectly using assign(), e.g.

```
assign("phi", (1 + sqrt(5)) / 2)
```

This second approach is helpful when the variable of interest only becomes known as the R code is being run, as is common in reading data files.

2.3.2.2 Variable Evaluation

Writing the name of a variable in an expression that is evaluated causes R to look up the value and use it appropriately,[6] e.g.

```
1 / phi
| [1] 0.618034
```

It is also possible to find a variable by name, using get(), e.g.

```
get("phi") - 1
| [1] 0.618034
```

(Aside: comparison of the previous two results reveals a key property of the golden ratio.)

---

[6] A subtle point is that R does not always look up the values of variables until they are needed. This is related to R concepts of "lazy evaluation" and "promises".

2.3.2.3 Variable Names

Readers with programming experience will find that R accepts common conventions for variable names. For example, a name may contain letters and numbers, the case of letters is significant, etc. However, there are important differences between R variable names and those in other languages.

In some languages, a period in a variable name has a special meaning, e.g. to refer to part of a multi-component object. This is not so in R, where the purpose of a period is both more varied and more confusing (see, e.g., Bååth 2012). In some cases, R treats a period in a variable name just like any other character, e.g. in

```
one.half <- 1 / 2
```

it provides a visual cue that two words are linked. However, in some circumstances, periods have a special meaning relating to the generic functions, a somewhat complex topic to be dealt with in Sect. 2.3.11.6. In the interest of clarity, many R users avoid periods in names, except for generic functions. That raises a question of how best to provide a visual cue of word linkage. A style that is popular in some other languages is to use underlines, e.g.

```
three_quarters <- 3 / 4
```

However, underline was once used to indicate assignment in R (a convention borrowed from S), and some text editors automatically expand this character to <-. To avoid confusion, the author uses "camel case" notation, in which a case switch is used to separate words, e.g.

```
fourFifths <- 4 / 5
```

This notation is used in the oce package and in other packages, including the popular Bioconductor system for genomic analysis.

Caution R provides great freedom in variable names. For example, assign() permits a blank character as a variable name, and get() recovers its value without complaint. Needless to say, such tricks are best avoided.

Caution R does not have read-only variables. For example, even though pi is automatically defined by R to be π, it is valid to write pi <- 3.41, and doing so can lead to errors that may be difficult to find later.

2.3.2.4 Variable Scope

Variables in R come into existence when they are assigned a value. Those values are accessible only within what is called the variable "scope". For example, variables created within a function are destroyed when the function returns. This is important, because it means that calling an R function tends not to have side effects on any variables in the enclosing scope. This lack of side effects is an aspect of so-called functional programming, in which a function affects its environment only through its return value.

Sometimes, however, it seems that side effects are the best solution to a coding problem, and the `<<-` operator may be used then, to widen the scope of assignment. See, e.g., Gentleman and Ihaka (2000) and Wickham (2014) for further discussion.

2.3.3 Basic Storage Types

2.3.3.1 Numerical Types

Integer numbers are denoted with suffix "L" (for "long", the size of integer storage in R), e.g.

```
five <- 5L
```

which is revealed as an integer by `storage.mode()`

```
storage.mode(five)
[1] "integer"
```

As noted previously, sequences of numbers can be generated with ":"

```
1:3
[1] 1 2 3
```

and more general increments are handled with `seq()`, e.g.

```
byTwo <- seq(10L, 20L, 2L)
```

produces the sequence 10, 12, ..., 20 as integers (but the returned vector would be of the "double" storage mode if the third argument were not an integer, or if other argument values required floating-point representation). Integers that map to the contents of the sequence can be produced with `seq_along()`

```
seq_along(byTwo)
[1] 1 2 3 4 5 6
```

and such mappings are often helpful in working through datasets, such as the stations with an oceanographic section, or the levels within a given station.

Floating-point numbers are indicated with decimal points or exponents

```
twoPi <- 8 * atan2(1, 1)
avogadro <- 6.02e23
```

Complex numbers are denoted with suffix "i" on the imaginary part, e.g.

```
x <- 1 + 2i
```

The real and imaginary parts of a complex number are recovered with `Re()` and `Im()`, and related functions perform other requisite tasks. Many R functions accept complex numbers as arguments, e.g.

```
sqrt(1i)
[1] 0.7071068+0.7071068i

exp(pi * 1i)
[1] -1+0i
```

Unfortunately, R does not provide small storage types, such as the 2-byte integers that are used to save space in several acoustic Doppler and satellite data formats. If memory is sufficient, it may be sensible to promote such data to 8-byte integers or floating-point values in R. However, for large datasets it is better to glue together pairs of single-byte elements, as is done in the `oce` package (Chap. 3).

2.3.3.2 Logical Type

R uses TRUE and FALSE to denote logical values, e.g.

```
waterIsWet <- TRUE
```

Logical negation is achieved by putting `!` to the left of a logical quantity. Writing `|` between two logical quantities yields "or", while `&` yields "and." Note that these operators produce vectors when applied to vectors, while the related operators `||` and `&&` each produce single-valued results.

A common way to construct logical values is through comparison, e.g.

```
x <- seq(-3, 3)
x < 0
 [1]  TRUE  TRUE  TRUE FALSE FALSE FALSE FALSE
```

shows how the less-than operator is used; similar operators include `<=`, `>`, `>=`, `==`, and `!=`. (See Exercise 2.7 for notes regarding the use of `==`.) A common use of logical values is for subsetting, e.g.

```
mean(x[x > 0])  # mean of positive values
 [1]  2
```

(This is equivalent to `mean(subset(x, x > 0))`, which is a clearer expression to some users.)

The indexing method also works for assignment, e.g.

```
x[x <= 0] <- NA
mean(x, na.rm=TRUE)
 [1]  2
```

where `mean()` has been given an argument to ignore missing values.

Caution Although TRUE and FALSE can be abbreviated T and F, this is a poor idea because it leads to confusion, especially in oceanographic work, where "T" is a common abbreviation for a temperature.

Exercise 2.7 Use `==` to find your computer's precision, i.e. the smallest resolvable difference between floating-point values. (See page 190 for a solution.)

Exercise 2.8 Explain why `all.equal()` is good way to compare floating-point values. (See page 190 for a solution.)

2.3.3.3 Textual (Character) Type

Text strings may be enclosed in single or double quotation marks, and one nested in the other is taken as a literal, e.g.

```
cat("Henry ('Hank') Stommel was a smart man.")
Henry ('Hank') Stommel was a smart man.
```

Strings may be pasted together with paste()

```
paste("Stommel", "(1948)", "is a classic.")
[1] "Stommel (1948) is a classic."
```

Substrings may be extracted with substr()

```
filename <- "atlantic.dat"
substr(filename, 1, 8)
[1] "atlantic"
```

and nchar() gives the number of characters in a string

```
substr(filename, nchar(filename)-2, nchar(filename))
[1] "dat"
```

Strings may be split into components with strsplit()

```
strsplit("Stommel-Arons-Faller", split="-")
[[1]]
[1] "Stommel" "Arons"   "Faller"
```

the result of which is a "list," discussed in Sect. 2.3.6. (Splitting is a very useful operation, for many tasks.)

R provides a variety of functions for altering strings, including sub(), which replaces the first matched substring

```
sub("a", "A", "atlantic")
[1] "Atlantic"
```

and gsub(), which (by default) replaces all occurrences

```
gsub("a", "A", "atlantic")
[1] "AtlAntic"
```

Various text encoding schemes may be used for strings, including ASCII, UTF-8, and "byte" forms. People requiring accents in strings probably know how to enter them with key combinations, e.g. the "ä" in "Väisälä" may be obtained with option-u a on some systems. Such characters may also be entered with a coding system known as ISO/IEC 8859-1. In this, the code for "ä" is the hexadecimal sequence E4. Such sequences may be entered into R strings by prefacing them with the two-character code \x and setting the encoding with Encoding(), e.g.

```
N <- "V\xE4is\xE4l\xE4"
Encoding(N) <- "latin1"
N
[1] "Väisälä"
```

Setting the encoding to "bytes" can guide the construction of regular expressions, but this can be a tricky business, depending on the encoding used to input data, etc.

It can be wise to stick with one encoding scheme. In oce, that scheme is (usually) UTF-8. To see how this works, consider reading CTD files created by Seabird software. If the software is set up to save σ_θ in CNV files, then there will be a column named sigma-é00. Then the relevant header line can be isolated in UTF-8 and latin-1 formats with, e.g.

```
f <- system.file("extdata", "d201211_0011.cnv",
package="oce")
readLines(f, encoding="latin1")[54]
```
> [1] "# name 22 = sigma-é00: Density [sigma-theta,
> Kg/m^3]"

```
readLines(f, encoding="UTF-8")[54]
```
> [1] "# name 22 = sigma-\xe900: Density [sigma-theta,
> Kg/m^3]"

Although the first form has the advantage of displaying the line as a text editor might, the UTF-8 form may be more convenient for programming, e.g. `read.ctd.cnv()` finds such entries in headers with

```
n <- gsub("^# name [0-9][0-9]* = (.*):.*$",
          "\\1", h, ignore.case=TRUE, useBytes=TRUE)
if (1==length(grep("^sigma-\xe9[0-9]{2}$", n,
useBytes=TRUE)))
```

Readers with programming experience will have no trouble reading regular expressions in the previous lines of code. Others might benefit from a brief sketch (consulting the R help system for details). Consider the strings `"Atlantic"` and `"Pacific"`, which may be joined into a vector with `c()`. To see which contains the letter `"t"`, write

```
grep("t", c("Atlantic", "Pacific"))
```
> [1] 1

If no element had contained the pattern, `grep()` would have returned a zero-length vector, so the expression `0 < length(grep(p, x))` tests whether any element of vector x contains pattern p.

Special characters can be inserted into patterns, achieving useful outcomes. The character `"^"` stands for the start of a string, and `"$"` stands for the end of the string, as in the CTD example above. A period can stand for any character. An asterisk after a pattern indicates that the pattern may appear zero or more times. Alternative characters are enclosed square brackets, and a dash can be used to create a sequence. Substrings enclosed within parentheses can be reused later, as is done in the `gsub()` example, where n is assigned the value of the first (and only) marked substring.

Special difficulties arise with human-entered text, because typos are common, and users in different locales might spell differently. R helps in such cases with fuzzy string matching provided by `agrep()` and `adist()`.

Exercise 2.9 A directory contains Biosonics echosounder files, with names indicating start times, with four digits for year, two for month and two for day, followed by an underline and then two digits for hour, two for minute, and two for second, ending with `.dt4`. Use `grep()` to isolate data starting between 1100 h and 1500 h on June 28th, 2008. (See page 190 for a solution.)

2.3.3.4 Binary (Raw) Type

R has a byte-level type that it designates as "raw." Positive integers in the range from 0 to 255 can be converted to this form with as.raw(), e.g.

```
as.raw(1:16)
  [1] 01 02 03 04 05 06 07 08 09 0a 0b 0c 0d 0e 0f 10
```

and as.raw() also handles floating-point values in the allowed range, by first coercing its input to integer form. Raw constants follow a C-like notation, e.g. 0x0f or 0x0F for $0000\,1111_2$, and C notation is also used for binary operations, with the conventional operators |, & and !.

A good way to handle binary formats is to use readBin() to read the whole file into a raw vector, and then to work though that vector with readBin() on smaller chunks of data. Readers who have struggled with files in which the endianness shifts from entry to entry will appreciate the fact that readBin() has an argument specifying the endian nature of the item being decoded.

2.3.3.5 Time Types

A simple approach to dealing with time in R is to treat it as a numerical value that has meaning to the user, but not to R. For example, a numerical model might output time in days or years, with the expectation being that an analyst will take the unit into account when making calculations and plotting results. However, a more systematic approach is required for data.

Decoding times recorded in notebooks can require answering questions such as whether 2018/2/10 is early in the year or late in it, whether 8 o'clock is in the morning or the evening, and when (or whether) daylight-savings time commenced in a given year, in a given jurisdiction ... and realizing that such answers might not hold across pages, even those written by a single person.

Some details are more universal, and R handles them well by itself. These include leap years and leap seconds, along with the thornier matter of timezones. (The IANA timezone database[7] spans over 20,000 lines.) Readers interested in such things should consult Ripley and Hornik (2001) for an introduction to how R handles dates and times.

R has two schemes for representing time types. In one, time is represented by a numerical value representing the year, another representing the month, etc. This scheme is used by as.POSIXlt(), which converts text strings into times. In the second scheme, used by the analogous function as.POSIXct(), time is represented by a single numerical value that measures the interval since a reference time. Both schemes have additional information on timezone, etc.

Conversion between these two schemes is trivial, so there is little need to discuss both. The single-number scheme will be the focus here. Time objects created by

---

[7] www.iana.org/time-zones.

as.POSIXct() store the number of seconds since the start of the year 1970.[8] Prior times have negative values, and fractional seconds are handled by the use of floating-point storage.

For example, one minute past the origin time is

```
t0 <- as.POSIXct("1970-01-01 00:01:00", tz="UTC")
```

This may be displayed as a text string or a numeric value

```
t0
```
```
| [1] "1970-01-01 00:01:00 UTC"
```
```
as.numeric(t0)
```
```
| [1] 60
```

Note the use of tz to set the timezone. If tz is not given, R will use a local time zone. This default can lead to highly undesirable results, such as an R program producing different results when run in different regions, so a tz value should always be supplied. However, even this is problematic, because of ambiguities in timezone notation. For example, AST means Atlantic Standard Time to the author, but it might mean Alaskan Standard Time, or Afghanistan Standard Time, to a reader. A good solution is to specify timezones by region and city, e.g.

```
as.numeric(as.POSIXct("1970-1-1 00:00:00",
                      tz="America/Halifax"))
```
```
| [1] 14400
```

shows that local standard time in the author's city is 4h "behind" UTC. (Use help(timezones) to learn more about timezones.)

Handling alternative representations of times is simplified with the format argument, e.g.

```
as.POSIXct("Jan 1, 1970 00:01:00", tz="UTC",
           format="%b %d, %Y %H:%M:%S")
```
```
| [1] "1970-01-01 00:01:00 UTC"
```

handles a common format used in non-technical writing; note that %b is an abbreviated month name, %d is decimal day, %Y is year including century, %H is hour, %M is minute, and %S is second. See the documentation for strptime() for the details of the coding scheme, which includes some standardized forms, e.g.

```
as.POSIXct("1917-12-06 09:04:35", format="%F %T",
           tz="America/Halifax")
```
```
| [1] "1917-12-06 09:04:35 AST"
```

expresses the time of the Halifax explosion in ISO 8601 format.

Using as.POSIXlt() is very similar to this. Working with numerical values for year, month, etc., is also easy, with ISOdatetime().

All of the above relates to the base package. The lubridate package (Grolemund and Wickham 2011) also has functions that parse common formats, and e.g.

---

[8] Alternative time origins may be specified to as.POSIXlt(), and this can be helpful in working with times represented in other systems such as SPSS and SAS.

```
library(lubridate)
ymd("1970 Jan 1")
[1] "1970-01-01"

ymd("1970-01-01")
[1] "1970-01-01"
```

demonstrates that it is quite adept at decoding formats.

Oceanographers use a wide variety of numerical schemes for time, so oce provides numberAsPOSIXct() for inferring times from the Unix, Matlab, SAS, SPSS, GPS and Argo numerical schemes.

The above has dealt with single values, but of course R also handles vectors of times. For example, sequences of times may be created with seq()

```
seq(t0, by="1 min", length.out=2)
[1] "1970-01-01 00:01:00 UTC" "1970-01-01 00:02:00 UTC"
```

or simply by adding a sequence of numerical values to a time, e.g.

```
t0 + 60 * seq(0, 1)
[1] "1970-01-01 00:01:00 UTC" "1970-01-01 00:02:00 UTC"
```

Despite the strong support for time types in R, there are many ways to get into trouble if care is not taken. A few hints may help.

1. Always set the timezone, ideally to tz="UTC".
2. Matlab users should note that Julian days start at 0 in R, e.g.

   ```
   as.numeric(julian(as.POSIXct("1970-1-1", tz="UTC")))
   [1] 0
   ```
3. Be careful when assembling multiple times with c(), e.g.

   ```
   as.POSIXct(c("1970-1-1", "1970-1-1 0:0:1"))
   [1] "1970-01-01 AST" "1970-01-01 AST"
   as.POSIXct(c("1970-1-1 0:0:0", "1970-1-1 0:0:1"))
   [1] "1970-01-01 00:00:00 AST" "1970-01-01 00:00:01 AST"
   ```

   reveals that R selects the printing format based on the first element.
4. Be aware that format() and strftime() handle unspecified timezones differently, e.g.

   ```
   t0 <- as.POSIXct("1970-01-01 00:01:00", tz="UTC")
   format(t0)
   [1] "1970-01-01 00:01:00"
   strftime(t0)
   [1] "1969-12-31 20:01:00"
   ```

   (Note that setting tz="UTC" in the format() and strftime() calls makes them yield identical results.)
5. Limit your choice of functions, and study the documentation well, to avoid surprises such as the one just mentioned.
6. Try using attr(), attributes() and as.numeric() to find the roots of any problems that may arise.

2.3.3.6 Missing Values and Other Special Values

Conventional notation is used for numerically problematic values in R, e.g.

```
c(1/0, -1/0, asin(3))
[1]  Inf -Inf  NaN
```

R has a special code for missing values, such as might result from instrument malfunction. These are indicated with NA, which could be read as "not appropriate." A missing value can take the place of most R items, e.g.

```
c(1, 2, NA, 4)
[1]  1   2 NA   4
```

```
c("inshore", NA, "offshore")
[1] "inshore"  NA          "offshore"
```

The provision of missing values in R reveals that it is a language that grew from the demands of research. Other systems such as Matlab reuse NaN or some other code for a missing value, blurring meaning. Many R functions detect missing values and offer ways to control how they are interpreted, e.g.

```
mean(c(1, NA, 2))
[1] NA
```

```
mean(c(1, NA, 2), na.rm=TRUE)
[1] 1.5
```

There are also functions for selecting data, e.g.

```
mean(na.omit(c(1, NA, 2)))
[1] 1.5
```

In many cases, simply omitting missing data will be sufficient, but sometimes this is not an option, e.g. skipping data in a time series will yield problems with computing spectral properties. An entry to the general literature about handling missing data is provided by Horton and Kleinman (2007). Readers can also find guidance in their own branches of the oceanographic literature, with treatments often being keyed to instrument type, e.g. Sect. 5.9.2.2 presents ideas for dealing with acoustic-Doppler velocimeter (ADV) data.

R uses the symbol NULL to indicate an extant but empty quantity. This is sometimes used in function arguments. A common use in processing is in the item-by-item construction of vectors, as explained in Sect. 2.3.4.

R provides several functions for testing whether numbers are problematic, e.g. is.nan() tests for NaN, is.infinite() tests for Inf, and is.null() tests for NULL values. In many cases, the best test is is.finite(), which returns TRUE only if the argument is not Inf, not NaN, and not NA.

Caution Matlab data files tend to use NaN for missing value, so that Matlab files that are converted to R need an extra step, e.g.

```
x <- readMat("x.mat")
x[is.nan(x)] <- NA
```

2.3.4   Vectors

Vectors hold sequences of values, and e.g.

```
x <- 3
is.vector(x)
[1] TRUE
```

reveals that even single values are vectors.

A vector can contain entries of any atomic mode, meaning "logical", "integer", "numeric", "complex", "character" or "raw", or of either "expression" or "list" mode. However, a vector cannot contain an admixture of these modes. Applications requiring the grouping of items of dissimilar modes should use lists (Sect. 2.3.6), instead of vectors.

A few examples will suffice to show how to use vectors in practical work. A vector of numerical values may be constructed in a number of ways, e.g. with the colon operator ": " for integers

```
threeStooges <- 1:3
```

with a sequence function for more general values

```
thirds <- seq(0, 1, length.out=4)
```

with the repeat function

```
fourScore <- rep(20, 4)
```

and with the collection function

```
irrational <- c(pi, exp(1))
```

The last of these methods also works for strings

```
stoogeNames <- c("Larry", "Curly", "Moe")
```

Vector elements are accessed with a square-bracket " [" notation patterned on C, although R indexes the first element at 1, not 0 as in C, e.g.

```
stoogeNames[2]
[1] "Curly"
```

and multiple elements may be accessed at the same time, by providing a vector of indices within the square brackets, e.g.

```
stoogeNames[c(1, 3)]
[1] "Larry" "Moe"
```

A boolean vector may also be used to access elements. For example, the most colourful of the seven seas of Medieval literature are

```
seas <- c("Mediterranean", "Adriatic", "Arabian",
          "Black", "Caspian", "Persian", "Red")
seas[c(FALSE, FALSE, FALSE, TRUE, FALSE, FALSE, TRUE)]
[1] "Black" "Red"
```

In the above examples, the vector contents are specified in code, so R can set up storage before assigning values. However, in some cases, vector length is discovered by data inspection or user instruction. If the requisite vector length is known before determining contents, storage may be allocated with e.g.

```
depths <- vector("numeric", 100)
```
with values entered later. If length is not known in advance, a vector can be constructed incrementally (less efficiently), starting with an empty vector
```
lengths <- NULL
```
and then, as each new length L is found, append it to the list, e.g.
```
lengths <- c(lengths, L)
```
Another approach is to assign past the end of the vector, e.g.
```
lengths[2] <- 4
```
appends an additional item to the vector. Starting in 2017, R developed a new scheme for memory allocation beyond the end of a vector, yielding great improvements to the processing speed of this method, making it preferable to the `c()` method in many cases.

Caution Binary operations between vectors follow a recycling rule that permits combination even if the vectors are of unequal length. This works by cycling through the elements of the shorter vector as needed, e.g. in
```
1:6 + c(1, 0)
[1] 2 2 4 4 6 6
```
the second vector is expanded to `c(0, 1, 0, 1, 0)` before the addition. This behaviour is very helpful, but can be surprising to programmers coming from languages that disallow operating on mismatched objects.

Exercise 2.10 Use `floor()` to select even integers from a vector. (See page 190 for a solution.)

2.3.5 Arrays and Matrices

As with vectors, the concept of a matrix should be familiar to most readers. A simple interpretation is a grid of values, as one might construct by writing a number (or string, etc.) in the boxes on a square-ruled sheet of paper. An array is a more general item that can have higher dimensions. For example, a gridded sea-level field $\eta = \eta(x, y)$ might be stored in a matrix, while a 3D `array` would suit a gridded temperature field $T = T(x, y, z)$.

As with vectors, arrays are fairly flexible in terms of their contents, but any given array can contain only one type, and that type must be single-valued (i.e. it is not possible to store an array in the cell of another array).

To see if an item is a matrix, use `is.matrix()`, and to see whether it is an array, use `is.array()`. Coercion rules usually ensure that arrays alter themselves if the type of an element alters, e.g. if a floating-point number is inserted into a matrix of integers, the rest of the numbers are converted to floating point values.

The matrix() function can construct matrices from vectors, e.g.

```
m <- matrix(1:6, nrow=2)
m
      [,1] [,2] [,3]
[1,]     1    3    5
[2,]     2    4    6
```

shows how to fill a matrix by columns; use byrow=TRUE to fill by rows. Matrices can also be created by combining columns with cbind() or rows with rbind().

2.3.5.1   Algebra

Matrix operations may be carried out in an element-by-element manner with operators such as +, *, etc. For example, each cell in A*B is the product of corresponding cells in A and B. (Note that a recycling rule applies to matrices, as for vectors.) Matrix multiplication is denoted by the %*% operator, and the usual linear-algebra rules control whether A%*%B can succeed, depending on the matrix dimensions. Those dimensions may be recovered or set with dim(). Other common matrix operations include: transpose with t(), determinant calculation with det(), singular value decomposition with svd(), QR decomposition with qr(), eigenanalysis with eigen(), inversion with solve(), generalized matrix inversion with ginv() from the MASS package, etc.

2.3.5.2   Indexing and Subsetting

Subsets of matrices can be extracted with [, e.g. with m as defined previously,

```
m[1,]
[1] 1 3 5
```

shows the first row, and

```
m[-1,]
[1] 2 4 6
```

shows the results of deleting that row.

The labelling in the results shown above indicates that R has converted the column and row matrices into vectors. This behaviour can be disabled by supplying the drop argument, e.g.

```
m[1,,drop=FALSE]
      [,1] [,2] [,3]
[1,]     1    3    5
```

Note the need for two commas here, because the second argument to [is an index; see help("[") for more.

Caution The R conversion to vectors should be kept in mind when converting matrix-oriented code from other languages.

2.3.5.3 Reshaping

In addition to reporting the dimensions of an item, `dim()` can be used on the left-hand side of an assignment expression, to set the dimension, e.g.

```
dim(m) <- c(1, 6)
m
       [,1] [,2] [,3] [,4] [,5] [,6]
[1,]     1    2    3    4    5    6
```

This reveals that matrices and arrays are stored in column order, which proves convenient when connecting R to C, C++ or Fortran.

2.3.5.4 Storage

R stores matrices and arrays as vectors, saving dimensions with attributes

```
attributes(m)
$dim
[1] 1 6
```

The dollar sign in the result is an indication that the attributes are stored in an item named `dim` within a list. (See Sect. 2.3.6 for more on lists.)

2.3.5.5 Example: A Rotation Matrix

Suppose moorings are placed at locations drawn in Fig. 2.3 (left) with

```
E <- seq(0, 0.5, 0.1)
N <- seq(0, 0.5, 0.1)
plot(E, N, xlim=c(0,1), ylim=c(0,1), asp=1)
```

and that these locations are to be expressed in an xy coordinate system rotated $45°$ anti-clockwise of geographic. This is expressed with rotation matrix

$$\mathbf{R} = \begin{bmatrix} \cos\theta & \sin\theta \\ -\sin\theta & \cos\theta \end{bmatrix} \tag{2.2}$$

Fig. 2.3 Rotation from E-N coordinate system to x-y coordinate system.

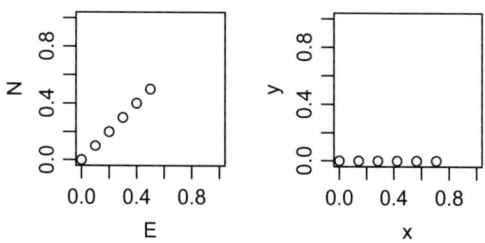

with $\theta = 45°$, which may be expressed as

```
theta <- 45 * pi / 180
S <- sin(theta)
C <- cos(theta)
R <- matrix(c(C, S, -S, C), byrow=TRUE, nrow=2)
```

so that matrix multiplication yields x and y in columns

```
xy <- R %*% rbind(E, N)
plot(xy[1,], xy[2,], xlim=c(0,1), ylim=c(0,1), asp=1,
     xlab="x", ylab="y")
```

as in the right panel of Fig. 2.3.

Exercise 2.11 Write a function to find the indices of the maximal value of a matrix. (See page 191 for a solution.)

2.3.6 Lists

The restriction that any given vector, matrix or array can hold just one data type yields efficiencies, but some programming situations call for collections of disparate items. In R, such collections are called lists, and they may be constructed with `list()`, e.g.

```
hex <- list(name="six", value=6, constituents=c(2,3))
```

or with `as.list()`. If the list items have names, they may be retrieved with "$" extraction operator or "[[" (this second operator being overloaded to look within oce objects; see Sect. 3.3).

```
hex$name
[1] "six"
hex[["value"]]
[1] 6
```

Regardless of whether list items are named, they may be also recovered by numerical index, with single-bracket notation

```
hex[3]
$constituents
[1] 2 3
```

retaining item names and double-bracket notation

```
hex[[3]]
[1] 2 3
```

discarding names.

As will be explained in Chap. 3, the oce package stores data as lists, because so many oceanographic data combine vectors with matrices, raw bytes with decimal numbers, etc.

Exercise 2.12 Show how to access a list within a list. (See page 191 for a solution.)

2.3.7 Factors

Factors designate categorical data, and they have wide applications to oceanographic work, for such things as species names, etc. Factors simplify data analysis because many R functions handle them in useful ways. The ideas may be illustrated with air-sea drag data reported by Garratt (1977).

```
data(drag, package="ocedata")
```

The values of the drag coefficient C_D stored as Cd within this dataset were inferred from either a "profile" method or an "eddy correlation" method. A factor is a natural way to store this information, as is revealed with

```
str(drag) # results not shown
```

The list of categories is provided by levels()

```
levels(drag$method)
[1] "profile" "eddy"
```

while as.numeric() reveals the method used for each datum

```
as.numeric(drag$method)
 [1] 1 1 1 1 1 1 1 1 1 1 1 1 1 2 2 2 2 2 2 2 2 2
     2 2 2 2
[29] 2 2 2
```

These numerical values are convenient for coding the data with symbols, as in the left panel of Fig. 2.4.

```
code <- as.numeric(drag$method)
levels <- levels(drag$method)
plot(drag$U, drag$Cd, pch=code,
     xlab="Wind Speed [m/s]", ylab=expression(C[D]))
legend("topleft", pch=1:length(levels), legend=levels)
```

As an example of R plotting functions that handle factors in special ways,

```
boxplot(Cd ~ method, data=drag, notch=TRUE)
```

yields the box plots shown in right panel of Fig. 2.4 (see Tukey (1977) for more on box plots and other methods of exploratory data analysis).

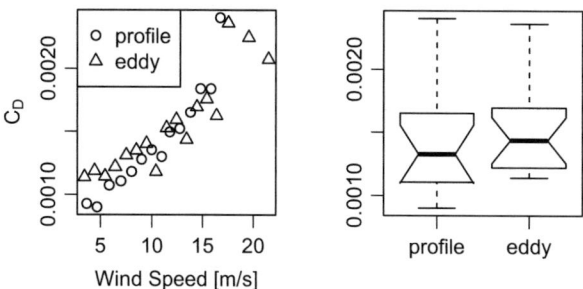

Fig. 2.4 Left: Symbol coding with factors. Right: boxplot coded with factors

Factors are also given special consideration by several non-graphical functions, and this lowers the effort of some everyday tasks, such as regression and the analysis of variance (Sect. 2.5.6).

More generally, factors provide a simple way to subdivide datasets into components to be processed individually, somewhat akin to the map-reduce method (Lämmel 2008) popularized by Google. For example,

```
dragSplit <- split(drag, drag$method)
```

splits `drag` into a list with elements named `profile` and `eddy`, making it easy to study differences between the subtypes, e.g. a *t* test

```
t.test(dragSplit$profile$U, dragSplit$eddy$U) $p.value
 [1]  0.4587837
```

might offer some insights.

For the `drag` data, the identification of `method` as a factor is straightforward. Factors can also be helpful with continuously varying data that can be subdivided meaningfully, as with the constructed O_2 profile[9] drawn as Fig. 2.5 with

```
p <- seq(0, 1000, 10)
O2 <- 280 - 1.7 * p * exp(-p / 200)
plot(O2, p, ylim=rev(range(p)), type="l")
```

Here, `p` is pressure in decibars and `O2` is oxygen concentration in μmol/kg. Figure 2.5 shows the profile constructed as follows.

If there were a need to find the thickness of the layer with O_2 concentration under 180 μmol/kg, a first step might be to create a factor

```
f<-factor(O2<180,levels=c(TRUE,FALSE),labels=c("Low",
  "High"))
```

(There is no requirement to supply `levels` and `labels`, but doing so reduces the chance of errors of interpretation.) To find the thickness of the layer in question, split the data

```
psplit <- split(p, f)
```

creating a list with elements named `Low` and `High`.

Fig. 2.5 Oxygen profile, with minimum region indicated

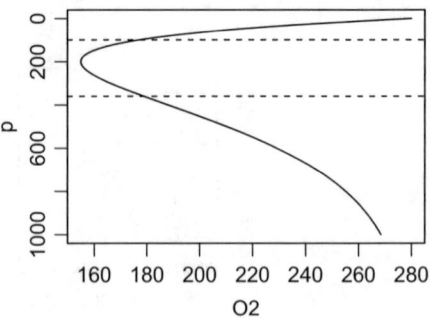

---

[9]This profile results from a nonlinear regression (Sect. 2.5.5.2) of the oxygen profile at station 112 of the `section` dataset in the `ocedata` package.

The pressure limits of the low-oxygen zone can be added to Fig. 2.5 with

```
abline(h=range(psplit$Low), lty=2)
```

Exercise 2.13 Use `factor()` and `split()` to identify the months in which Keeling CO_2 signal rises and falls. (See page 191 for a solution.)

2.3.8 Data Frames

Data frames are a complement to matrices and lists, and they are very important in R. Indeed, `help(data.frame)` states that data frames are "used as the fundamental data structure by most of R's modeling software."

One way to think of data frames is as matrices that may contain columns of different storage types. This extension can be very useful for general data, e.g. in listing information on two leaders of oceanography

```
leaders <- data.frame(person=c("Munk", "Stommel"),
                        born=c(1917, 1920))
```

The `print()` function displays an entire data frame

```
print(leaders)
      person born
1       Munk 1917
2  Stommel  1920
```

while `str()` provides an overview

```
str(leaders)
'data.frame':            2 obs. of  2 variables:
 $ person: Factor w/ 2 levels "Munk","Stommel": 1 2
 $ born  : num  1917 1920
```

that reveals that R created the first column as a factor.

Data frame columns may be selected in various ways:

```
leaders[1]      # or leaders["person"]
      person
1       Munk
2  Stommel
leaders[[1]]    # or leaders[["person"]]
[1] Munk     Stommel
Levels: Munk Stommel
leaders$person
[1] Munk     Stommel
Levels: Munk Stommel
```

The `$` style is convenient for interactive work, but the others are required for programs that need to use `names()` to find names at run time.

It is easy to modify data frames, e.g.

```
leaders$born<-c(as.Date("1917-10-19"),as.Date
("1920-09-27"))
```

alters the birth dates and

```
leaders$institution <- c("SIO", "WHOI")
```

adds a new column for workplace.

Exercise 2.14 Construct a data frame with column x containing numbers from 0 to 2π, and y containing $\sin x$. (See page 192 for a solution.)

Exercise 2.15 Append volume to the `oceans` dataset from the `ocedata` package. (See page 192 for a solution.)

Exercise 2.16 Suppose a data frame contains CTD data for a series of stations, with columns for salinity, temperature, pressure, and station ID. Use `split()` and `factor()` to create a list with one element per station. (See page 193 for a solution.)

2.3.9   Contingency Tables

Tabulation may be accomplished with functions `table()` or `tabulate()`. The first of these returns an object of class `"table"`, which is essentially an integer vector with names, while the second returns a integer without names.

A handy use of `table()` is to count missing values, e.g. the number of missing phosphate data in the `section` dataset from the `ocedata` package can be found as follows:

```
data(section, package="oce")
table(sapply(section[["phosphate"]], is.na))
FALSE   TRUE
 2817     24
```

where `sapply()` applies `is.na()` to all the PO_4 measurements in the section.

2.3.10   Conditional Evaluation

R provides conditional evaluation with `if` statements, e.g. the following checks whether the sun is above the horizon.

```
if (sunAzimuth > 0)
    cat("daytime\n")
```

In this case, a single action is to be performed. As in many other computing languages, multiple actions are enclosed with braces, e.g.

```
if (ocean == "Atlantic") {
    area <- 1e8 # km^3
    depth <- 4  # km
    cat("Ocean volume:", area * depth, "km^3\n")
}
```

It is also possible to specify a statement (or block of statements) to be executed if the tested condition is false, e.g.

```
if (depth < 100) cat("shelf\n") else cat("deep\n")
```

which also illustrates that blocks can be put on one line.

The `if` statement acts as an expression that returns a value, e.g.

```
sedimentType <- if (L < 62.5e-6) "mud" else "sand"
```

distinguishes between two sediment categories; nested `if`s might be used if samples might contain gravel, etc.

In the previous examples, only a single value was being tested. Multiple values can be handled with loops (Sect. 2.3.12), but it is usually better to use `ifelse()` to improve speed and clarity.

The first argument to `ifelse()` is a vector, matrix, etc., of logical values. The second is a corresponding vector (etc.) of values to be selected if a particular test value is TRUE, and the third contains values for FALSE conditions. For example, a matrix of depths

```
H <- matrix(c(10, 50, 90, 200), nrow=2)
```

can be categorized by domain with

```
ifelse(H < 100, "shelf", "deep")
     [,1]    [,2]
[1,] "shelf" "shelf"
[2,] "shelf" "deep"
```

2.3.11 Functions

2.3.11.1 Built-In Functions

R provides functions for common tasks related to plotting, statistics, and numerical analysis. Many specialized mathematical functions are provided as well, either in the base system or in packages. Some searching may be required to find the best package for a given task. For example, the formula for the perimeter of an ellipse involves the complete elliptic integral of the second kind, and this is available as `ellint_Ecomp()` in the `gsl` (GNU scientific library) package,[10] e.g. the perimeter of an ellipse with radii 1 and 2 is

---

[10]See http://www.gnu.org/software/gsl/ for more on GSL.

```
library(gsl)
a <- 2
b <- 1
4 * a * ellint_Ecomp(sqrt((a^2-b^2)/a^2))
[1] 9.688448
```

2.3.11.2   Defining Functions

The syntax for defining functions is similar to that for defining variables, using the assignment operator "<-". For example, ship speeds reported in knots can be converted to metres per second with a function defined as

```
knotToSI <- function(k) (1852/3600) * k
```

This assigns to the symbol knotToSI a function that takes a single argument named k. The expression following the list of arguments is the value to be returned by the function.

As with if statements, functions with more than one line need braces, e.g.

```
knotToSI <- function(k) {
    factor <- 1852/3600
    factor * k
}
```

The rule is that the last item evaluated in the function provides the return value. It is also possible to return a value specifically with return().

Of course, the purpose of this function is to work with numbers. Calling this with a non-numeric argument will generate an error. Good functions will check for erroneous argument values, and they will be flexible enough to handle a range of conditions. For example, knotToSI() might be extended to handle marine-telegraph specifications such as k="dead slow", by inserting a conditional that uses is.numeric() and adjusts k accordingly.

Function arguments may have default values, e.g.

```
knotToSI <- function(x, modern=TRUE)
    if (modern) x * (1852/3600) else x * (1853.248/3600)
```

permits the use of nautical miles as defined in the US prior to 1954, the year when the nation adopted the international standard.

Functions may check to see whether an argument had been supplied at call time, using missing(), e.g.

```
knotToSI <- function(x, modern=TRUE)
{
    if (missing(x))
        stop("must supply x")
    # rest of function as in previous examples
}
```

reports an error if nothing can be calculated; another choice might be

```
    if (missing(x))
        return(NA)
```

The argument list may contain ellipses (...) to indicate that there may be additional arguments that may be passed on to children, e.g.

```
indicateTheOcean <- function(ocean, ...)
    mtext(ocean, ...)
indicateTheOcean("Pacific", col="blue")
```

uses `mtext()` to draw text in a plot margin, using a blue colour. Importantly, `indicateTheOcean` does nothing related to colour; it merely provides the calling function with an opportunity to specify arguments such as `col` along to `mtext()`.

Many useful R functions have a large number of arguments, which would make it easy to make errors in calling the functions, but for the fact that R permits named arguments. This permits the skipping of arguments, and the specification of arguments out of order, e.g.

```
f <- function(x, y, z, u, v, w) ...
f(z=Z, w=W)
```

would make sense in situations not requiring x, y, u and v.

Exercise 2.17 Devise a function using `ifelse()` that returns the tangential velocity in a Rankine vortex. (See page 193 for a solution.)

2.3.11.3 Recursive Functions

Functions may be recursive, meaning that they can call themselves. Some algorithms are defined elegantly in such terms. For example, the greatest common denominator of two integers can be computed with

```
gcd <- function(a,b) if (b == 0) a else gcd(b,a %% b)
```

where the `%%` operator computes the remainder of division of two integers. However, as in other languages that permit recursion, there can be a computational penalty for mimicking elegant mathematical recursion in code (see Appendix E for some notes on handling computationally demanding tasks).

2.3.11.4 Functions as Arguments to Other Functions

Many important R functions take other functions as arguments. This scheme can be valuable, as may be illustrated with examples of three commonly used functions in this class: `uniroot()`, `optimize()` and `lapply()`.

First, consider the case of root-finding. Suppose the task is to find the temperature at which water of practical salinity 35 has density 1025 kg/m³ at atmospheric pressure. To do this, define the function

```
densityMismatch <- function(T)
    swRho(rep(35, length(T)), T, 0, eos="unesco") - 1025
```

which should have a root at the desired temperature.[11] A search interval for the root may be found with, e.g.

```
T <- seq(0, 30, length.out=100)
plot(T, densityMismatch(T), type="l")
abline(h=0, col=2)
```

which produces a graph (not shown) verifying that this function indeed has a zero in the interval $0°C < T < 30°C$, so

```
uniroot(densityMismatch, interval=c(0, 30))$root
| [1] 19.08284
```

provides the desired temperature.

Optimization provides a second example of functions calling functions. For the multivariate case, R provides `optim()`, plus related functions such as `nlm()` and `nls()`. There are subtle details to multivariate optimization, and so a better preliminary illustration is provided by the one-dimensional case, which is handled by `optimize()`. For example, suppose the goal is to find the temperature that yields maximum fresh-water density at sealevel pressure. With salinity pressure fixed, we may write a univariate function:

```
dens <- function(T)
    swRho(salinity=0,temperature=T,pressure=0,eos=
    "unesco")
```

so that the desired temperature may be computed with

```
optimize(dens, interval=c(0, 10), maximum=TRUE)$maximum
| [1] 3.980739
```

where the `maximum` value overrides the default, which is to seek a minimum.

Exercise 2.18 Use `uniroot()` and `coriolis()` from the oce package, to find the critical latitude at which the Coriolis parameter f matches the M2 tidal frequency (12.4206 hour period). (See page 193 for a solution.)

Exercise 2.19 Use `uniroot()` to create a function that calculates linear gravity wave speed as a function of period. (See page 193 for a solution.)

2.3.11.5   Function Closures

Function closures provide a mechanism for binding functions with parameters, making it easy to create suites of related functions that are made distinct by those parameters. Although this methodology may be unfamiliar to some readers, it is worth learning because it can improve code simplicity and reusability. The scheme uses functions that create other functions, e.g.

---

[11]Note the use of the UNESCO equation of state here; with the GSW equation, longitude and latitude would also have to be supplied; see Sect. 5.2.1 and Appendix D.

```
exponent <- function(p)
     function(x)  x^p
```
makes a function that creates functions for exponentiation:
```
cubeRoot <- exponent(1/3)
```
creates a cube-root function, called as `cubeRoot(8)`, for example.

Exercise 2.20 Create a function closure for individualized calibration of Seabird thermistors. (See page 194 for a solution.)

2.3.11.6 Generic Functions

Items in R have an attribute named `class`, which is revealed by `class()`
```
atl <- "Atlantic"
class(atl)
  [1] "character"
```
and changed with the same function
```
class(atl) <- "ocean"
class(atl)
  [1] "ocean"
```

A form of object orientation[12] is achieved in R via "generic" functions that are replaced by specialized functions according to the class of the first argument. The syntax is simple, with the specialized function being named as the desired generic function, followed by period and then the class name, e.g.
```
print.ocean <- function(x)
     cat("My favourite ocean is the", x, "\n")
```
defines a function to be used if `print()` is called with an `ocean` object.
```
print(atl)
  My favourite ocean is the Atlantic
```
Generic functions provide users with specialized functions without demanding that they know the individualized names of those functions, or even the classes of the objects under consideration. R provides specialized variants for such key functions as `print()`, `summary()` and `plot()`, and this greatly simplifies processing, as users tend to rely on the default of `plot(x)` doing something sensible, no matter what x may be.

A list of generic functions can be retrieved with, e.g., `methods(plot)`, and help on specialized functions is found by appending the class name, e.g. `help(plot.ts)` yields help on the time-series plot function.

---

[12]Readers who wish to learn more details of object orientation in R might start with Chambers (2008) or Wickham (2014).

2.3.11.7 Function Pipelines

In recent years, efforts have been made to implement an R analogy to Unix pipelines, with a notation for chaining function calls. For example, the mean sine of the integers from 1 to 10 may be computed conventionally in R with

```
mean(sin(1:10))
```
and the `magrittr` package lets this be written

```
library(magrittr)
1:10 %>% sin %>% mean
```
where `%>%` is a binary operator that takes the item on its left and supplies it as the first argument to the function on its right.

The scheme also works with user-generated functions, including anonymous function, which can use "." as a place-holder for the passed value, e.g. $\sum_0^{10} (1/2)^n$ becomes

```
1:10 %>% {(1/2)^.} %>% sum
```
```
[1] 0.9990234
```

Parentheses permit extra arguments to be supplied, e.g.

```
1:10 %>% sin %>% mean(trim=0.10)
```
uses a trimmed mean.

Some clarity is lent to complex operations by using line breaks, e.g.

```
1:10 %>%
    sin %>%
    mean(trim=0.10)
```
partly because of the space provided for comments on the individual steps.

2.3.11.8 Operators as Functions

In R, operators are functions. Thus, when the R parser encounters the division operator, it calls a function named "/", so that `1/2` is equivalent to

```
`/`(1, 2)
```
```
[1] 0.5
```

Readers with programming experience will see how this relates to the previous section, and might start using it for wider purposes. However, some care is required, e.g. the following shows how to turn addition into subtraction

```
`+` <- `-`
3 + 2
```
```
[1] 1
```
Note that the original meaning of + is recovered with

```
rm(`+`)
```

2.3.12 Loops

R has several styles of looping structures that provide for repeated calculation. The choice of structure is sometimes dictated by the problem at hand, and sometimes by personal style.

In a `for` loop, an index term is set to each value in a sequence, e.g.

```
for (i in 1:10)
    cat(2^i, ' ')
2   4   8   16   32   64   128   256   512   1024
```

Note that loops comprising more than one line require braces, just like conditional blocks with more than one line. In many cases, the iteration will be over the indices of items in a list or a vector, and a good way to handle this is to use `seq_along()` (as in Sect. 2.3.3.1) to find the indices; also, note that, e.g.

```
for (i in seq_along(x))
    print(x[i])
```

will not execute the loop if x is empty, whereas

```
for (i in 1:length(x))
    print(x[i])
```

will try to execute the loop contents with $i=1$ and then with $i=0$, perhaps surprising those who have not studied how ":" works. It pays to get in the habit of using `seq_along()` or its cousin, `seq_len()`. Also, if an index is not actually needed, it may be clearer to write, e.g.

```
for (n in seq(0, 5))
    cat(factorial(n), " ")
1   1   2   6   24   120
```

Although `for` is a natural way to loop over discrete cases, some methods are better expressed with a `while` loop, which repeats while a condition remains TRUE. For example, Heron's method for estimating square roots is

```
x <- 4 # number whose square root is desired
r <- 1 # first guess
while (abs(r^2 - x) > 0.01)            # tolerance 0.01
    r <- 0.5 * (r + x / r)
r
[1] 2.00061
```

A more basic loop is `repeat`, which never exits unless a `break` is executed.

```
x <- 0
repeat {
    cat(x, ' ')
    x <- x + 1
    if (x > 5)
        break
}
0   1   2   3   4   5
```

Actually, `break` can be used to break out of any type of loop, not just `repeat` loops. A relative is `next`, which causes a short-circuit that returns to the top of the loop, e.g.

```
for (x in seq(-1, 1)) {
    if (x < 0)
        next
    print(x)
}
[1]  0
[1]  1
```

Exercise 2.21 Write a loop that displays the values of items in the current workspace, using `ls()` and `get()`. (See page 194 for a solution.)

2.3.13 *Alternative to Loops*

Loops are not always desirable, e.g. the addition of two vectors with a loop

```
nx <- length(x)
z <- vector("numeric", length=nx)
for (i in 1:nx)
    z[i] <- x[i] + y[i]
```

is slower[13] and more difficult to understand than the non-looping form

```
z <- x + y
```

For this example, the looping form is also less general than the second form, because the latter handles matrices and arrays, not just vectors.

In addition to arithmetic cases such as the above, R offers a variety of ways to avoid loops. For example, `apply()` uses a given function for the elements of matrices or arrays, with

```
m <- matrix(1:8, nrow=2, byrow=TRUE)
m
       [,1] [,2] [,3] [,4]
[1,]    1    2    3    4
[2,]    5    6    7    8
apply(m, 1, sum)
[1] 10 26
```

showing how to sum along rows. (Use the second argument to 2 to sum across columns.) The third argument is any function with a single argument, e.g.

```
apply(m, 1, function(x) sum(x^2))
[1]   30 174
```

computes the sum of squares along rows.

---

[13]For more on performance issues, see Appendix E.

If the item under consideration is a list, then `lapply()` should be used, e.g.

```
l <- list(a=0:100, b=c(9,11))
lapply(l, mean)
```

(results not shown) computes the mean values of a and b.

There are several other functions in the "apply" family that are worth learning about. Readers familiar with Google's map-reduce method (Lämmel 2008) will see an analogy with the approach used in R. This topic is further discussed by Wickham (2011), both generally and in the context of his `plyr` and `dplyr` packages, which extend and simply the `apply` family of functions in base R.

Exercise 2.22 Extract velocity from the `oce` dataset adp, and plot distance-averaged beam-1 velocity versus time. (See page 194 for a solution.)

Exercise 2.23 Calculate and plot yearly average CO_2 data, using `lapply()`. (See page 195 for a solution.)

Exercise 2.24 Use a function in the `plyr` package to find minima and maxima of the data stored in `ctd[["data"]]`, a CTD station provided by the `oce` package. (See page 196 for a solution.)

2.4 Graphics

A great strength of R as a research tool is the simplicity and power of its graphics system. The use of generic functions (Sect. 2.3.11.6) ensures that applying `plot()` to differing data types produces results tailored to those types. For example, supplied with two columns, `plot()` produces an x-y plot. Supplied with a data frame, it creates a useful multi-panel graph that compares every column with every other column. Supplied with the results of a regression, `plot()` produces a set of deeply informative plots. In addition to such general things, many R packages provide plotting types, e.g. `oce` extends `plot()` to provide dozens of specialized oceanographic plots.

2.4.1 Scatter and Line Plots

Several examples of scatter plots and line plots having already been shown, more complicated examples may be of interest at this point in the text.

The base R system does not provide polar plots, but these can be constructed without difficulty. Consider the hourly wind speed and direction measurements held in the `buoy` dataset of the `ocedata` package.

```
data(buoy, package="ocedata")
```

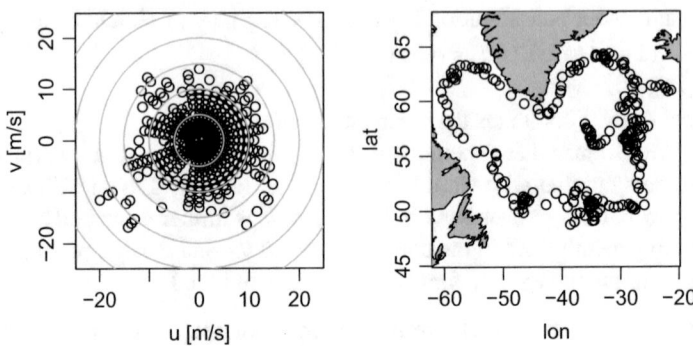

Fig. 2.6 Left: winds near Halifax. Right: Argo float trajectory

The data frame stores direction from which the wind blows, measured clockwise from true North, so the air velocities ($u > 0$ meaning flow to east, etc.) are computed with

```
theta <- (90 - buoy$direction) * pi / 180
u <- -buoy$wind*cos(theta)
v <- -buoy$wind*sin(theta)
```

and the left panel of Fig. 2.6 is created with

```
s <- c(-1, 1) * max(buoy$wind, na.rm=TRUE)
plot(u, v, xlab="u [m/s]", ylab="v [m/s]",
     xlim=s, ylim=s, asp=1)
for (ring in seq(5, 30, 5))
    lines(ring*cos(seq(0, 2*pi, pi/32)),
          ring*sin(seq(0, 2*pi, pi/32)), col="gray")
```

Note the use of xlim and ylim to centre the diagram, xlab and ylab to control labels, and asp to set the aspect ratio; these are all optional arguments that would likely be skipped in a quick plot.

As a second example, consider the argo dataset, representing Argo float measurements made about once per ten days. This is an oce object, so the location data can be extracted with the accessor operator, [[(Sect. 3.3)

```
data(argo, package="oce")
lat <- argo[["latitude"]]
lon <- argo[["longitude"]]
```

after which it is a simple matter to plot the float trajectory[14] with a coastline[15]

```
plot(lon, lat, asp=1/cos(pi*mean(range(lat))/180))
data(coastlineWorldMedium, package="ocedata")
cwlon <- coastlineWorldMedium[["longitude"]]
```

---

[14] See Sect. 5.7 for more on Argo floats.

[15] Several coastline resolutions are provided in the ocedata and oce packages.

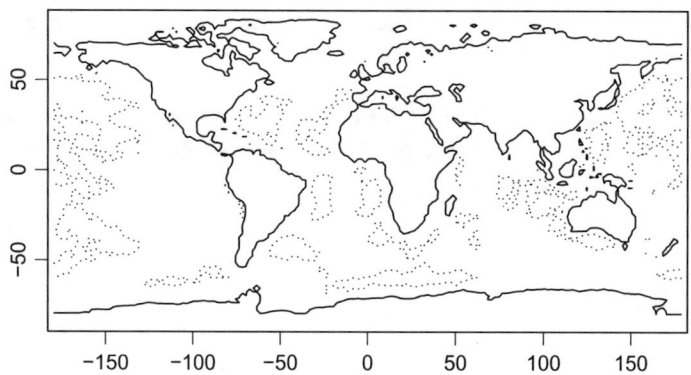

Fig. 2.7 Contoured world topography, showing coastline and 5 km isobath

```
cwlat <- coastlineWorldMedium[["latitude"]]
polygon(cwlon, cwlat, col="gray")
```

2.4.2 Contour Plots

The `ocedata` package provides a 2-degree resolution topographic dataset called `topo2`, a simple matrix of elevation in metres above mean sea level. The matrix may be contoured simply, with `contour(topo2)` yielding a serviceable diagram with auto-selected contour intervals and axes running from 0 to 1, but the following yields a more informative result (Fig. 2.7); note the use of `xaxs` and `yaxs` to prevent distracting space around the plot.

```
data(topo2, package="ocedata")
lon <- seq(-179.5, 178.5, length.out=180)# see ?topo2
lat <- seq(-89.5, 88.5, length.out=90)
contour(lon, lat, topo2, drawlabels=FALSE,
        levels=c(0,-5000),
        lty=c(1, 3), xaxs="i", yaxs="i", asp=1)
```

Caution Users familiar with Matlab contouring may find R contouring confusing. If the `topo2` dataset is saved to a file with `write.table()` (with `row.names` and `col.names` both FALSE) and loaded into Matlab with `load topo2.dat`, then the matrix will have to be transposed before contouring in Matlab. A simple mnemonic should make the R approach clear: the mathematical form $z(x, y)$ translates to the code form `z[ix, iy]`, where `ix` and `iy` are indices for the x and y grid directions.

Exercise 2.25 Reproduce Fig. 2.7 with axes labelled in geographical notation. (See page 196 for a solution.)

Exercise 2.26 Devise a wrapper function to handle reversed x or y values in contouring. (See page 196 for a solution.)

Exercise 2.27 Contour the formula for wind-chill temperature, $13.12 + 0.6215T - 11.37U^{0.16} + 0.3965TU^{0.16}$, as a function of air temperature, T, and wind speed, U. (See page 196 for a solution.)

2.4.3   Image Plots

The R function image() produces basic images, e.g.

    image(topo2)

yields the top panel of Fig. 2.8. There is no provision for a colour bar, but this can be created with imagep() in the oce package, e.g.

    imagep(topo2)

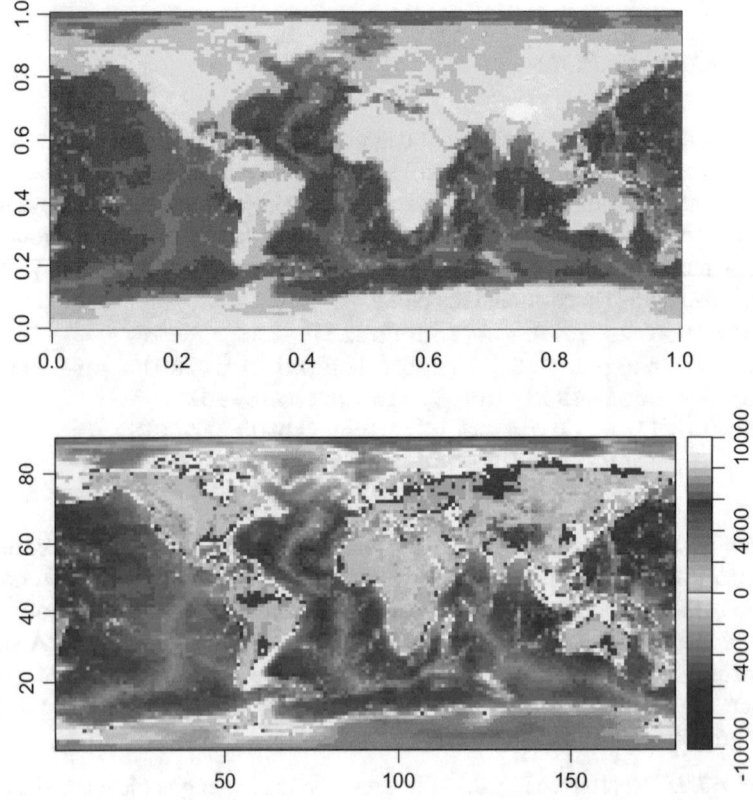

Fig. 2.8 Images produced with image(), top, and imagep(), bottom

produces an image (not shown) that is similar, apart from the different default colour scheme and the axis labelling.

Since the data represent topography, it makes sense to set a colour scale used in the literature for such fields. For example, the colour scale named "globe" within the Generic Mapping Tools software (Wessel et al. 2013) may be specified with

```
imagep(topo2, colormap=colormap(name="gmt_globe"))
```

which creates the bottom panel of Fig. 2.8. The `imagep()` axis labelling indicates the matrix dimensions, which may be more useful than the unit range shown by `image()`. Another practical consideration is that `imagep()` handles several special cases for its first argument, e.g. if it is a `topo` object as defined by the `oce` package, then `imagep()` will extract longitude and latitude, and use these on the axes.

2.4.4 Hexagon Binning

R offers several ways to show data density in a two-dimensional space, including `smoothScatter()` in the base `graphics` package (Exercise 2.28), and more sophisticated variants in other packages. The hexagon bin scheme, in the `hexbin` package, is worth illustrating because it is also popular in other computing systems (Carr 1991).

As an example, the distribution of salinity and temperature values in the `papa` dataset (from Ocean Weather Station P) may be displayed as in Fig. 2.9 with

```
library(hexbin)
data(papa, package="ocedata")
S <- as.vector(papa$salinity)
T <- as.vector(papa$temperature)
plot(hexbin(S, T, xbins=20))
```

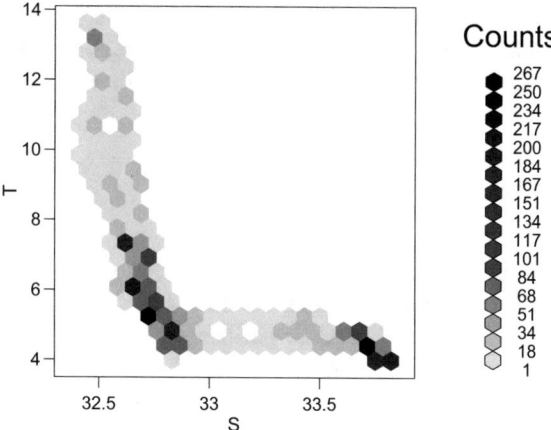

Fig. 2.9 Hexagon bin representation of data density in the papa dataset

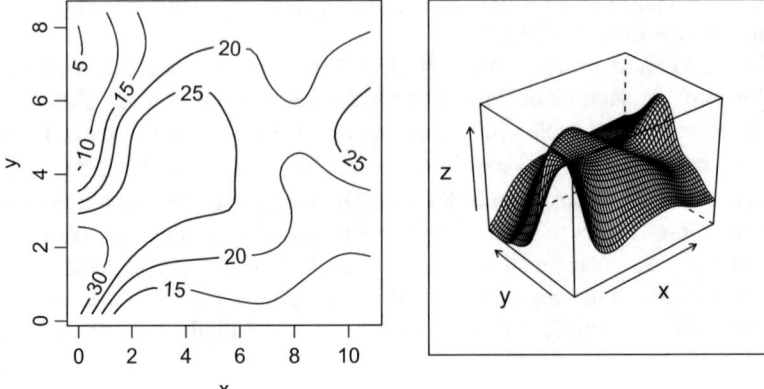

Fig. 2.10 Three-dimensional data represented with contours and a wireframe

2.4.5   Three-Dimensional Plots

The `rggobi` package provides interactive 3D plots, while the `lattice` package provides static ones. Such graphs being best suited to smooth functions, an example will be constructed using `interpBarnes()` to smooth the `wind` dataset.

The contour diagram in the left panel of Fig. 2.10 is created with

```
library(lattice)
data(wind, package="oce")
g <- interpBarnes(wind$x, wind$y, wind$z)
contour(g$xg, g$yg, g$zg, xlab="x", ylab="y", labcex=1)
```

and the 3D wireframe plot is added as the right panel with

```
wireframe(g$zg, xlab="x", ylab="y", zlab="z", cex=5)
```

Adjusting the viewing angle of wireframe plots can sometimes reveal features of interest, but often at the expense of obscuring other features. Ragged fields are a particular problem, e.g. `wireframe(topo2)` yields an uninformatively dense blob. It can be best to avoid eye-catching 3D plots, relying instead on more quantitative formats such as contour diagrams and images.

2.4.6   Time-Series Plots

The `ts` class handles time series data that are sampled at a constant rate. Constructing and plotting such things is easy. For example, the `giss` dataset (Hansen et al. 2010) from the `ocedata` package can be converted to a `ts` object and plotted as in the left panel of Fig. 2.11 with

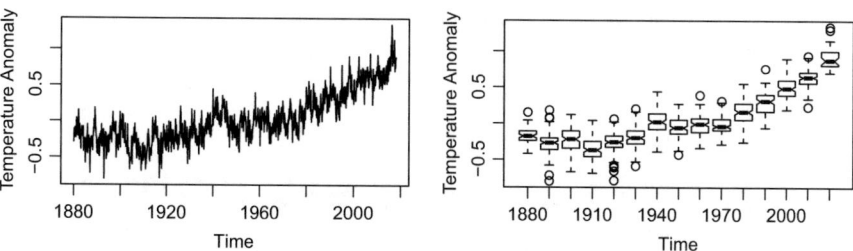

Fig. 2.11 GISS surface temperature anomaly illustrated with time-series and box plots

```
data(giss, package="ocedata")
Ta <- ts(giss$index, start=giss$year[1],
         frequency=1 / mean(diff(giss$year)))
plot(Ta, ylab="Temperature Anomaly")
```
(The automatic naming of the horizontal axis is a sign that the plot function is decoding time information stored in `attributes(Ta)`; such a scheme also facilitates spectral analysis, discussed later.)

2.4.7 Box Plots

The monthly `giss` signal being fairly ragged, an analyst might wish to construct statistical descriptions over years or decades. A box plot can be a good starting point for this sort of analysis (see, e.g., Tukey 1977). The `round()` function may be used to round to powers ten so that

```
decade <- round(time(Ta), -1)
```
yields an indicator of decade, and then

```
boxplot(Ta ~ decade, notch=TRUE,
        xlab="Time", ylab="Temperature Anomaly")
```
gives a box plot as in the right panel of Fig. 2.11. The `notch` setting yields a display of a type of confidence interval on the median, which is desirable in some applications.

2.4.8 Lagged Autocorrelation Plots

Lagged autocorrelation analysis offers further insights on the `giss` temporal variability, e.g.

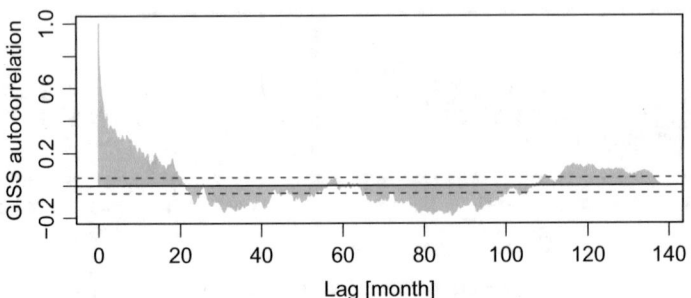

Fig. 2.12 Autocorrelation analysis of GISS surface temperature anomaly

```
Tad <- Ta - predict(lm(giss$index ~ giss$year))
acf(Tad, lag.max=length(Tad), col="gray",
    xlab="Lag [month]", ylab="GISS autocorrelation")
```
yields Fig. 2.12 for a detrended version of the data. (The maximum lag has been set to the length of the time series, because the default is too small to show the interannual variation.)

2.4.9 Histogram Plots

Histograms may be created with hist(), e.g. histograms of the velocity components in the adp dataset may be constructed with
```
data(adp, package="oce")
velo <- adp[["v"]]
eastward <- velo[,,1]
northward <- velo[,,2]
upward <- velo[,,3]
hist(eastward, main="")
hist(northward, main="")
hist(upward, main="")
```
yielding Fig. 2.13. The main argument to hist() prevents an unaesthetic label at the top of each panel. Another argument that is commonly set is breaks, which controls the bins of the histogram. In some cases, it can help to extract the return value from hist(), to construct other types of plots. For example, a cumulative histogram could be constructed with
```
eh <- hist(eastward, plot=FALSE)
plot(eh$mids, cumsum(eh$counts), type="s")
```
where plot=FALSE prevents hist() from plotting. The cumulative histogram, not shown here, uses "staircase" type to reveal the break points.

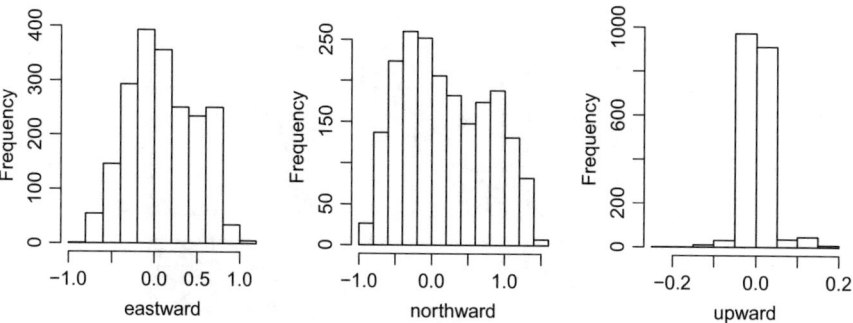

Fig. 2.13 Histograms of velocity measured with an Acoustic Doppler Current Profiler

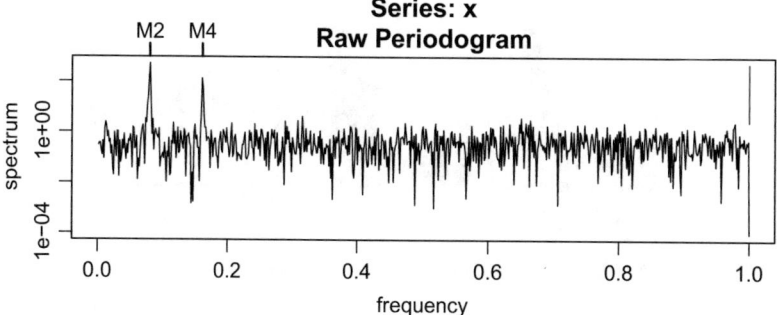

Fig. 2.14 Spectrum of an artificial time-series. Note that the vertical axis is logarithmic

2.4.10 Spectrum Plots

Spectral analysis is a complicated subject that will be discussed in some detail in Sect. 5.9.4.5. Here, the goal is just to introduce plotting. For example,

```
t <- seq(0, 30 * 24, 1/2)
f <- list(M2=0.0805114007, M4=0.1610228013) # cph
x <- 2*sin(f$M2*2*pi*t) + sin(f$M4*2*pi*t) + rnorm(t)
```

simulates a month of half-hourly sampled tidal data, which may be converted to a time series with

```
xts <- ts(x, frequency=2)
```

after which the spectrum shown in Fig. 2.14 may be constructed with

```
spectrum(xts)
rug(c(f$M2, f$M4), side=3, tcl=-0.5, lwd=2)
mtext(c("M2", "M4"), side=3, line=0.5, at=c(f$M2, f$M4))
```

The vertical line above the highest frequency is actually a cross indicating uncertainty in spectral value and bandwidth, but it looks like a line because the bandwidth is very narrow, without spectral smoothing (see Sect. 5.9.4.5).

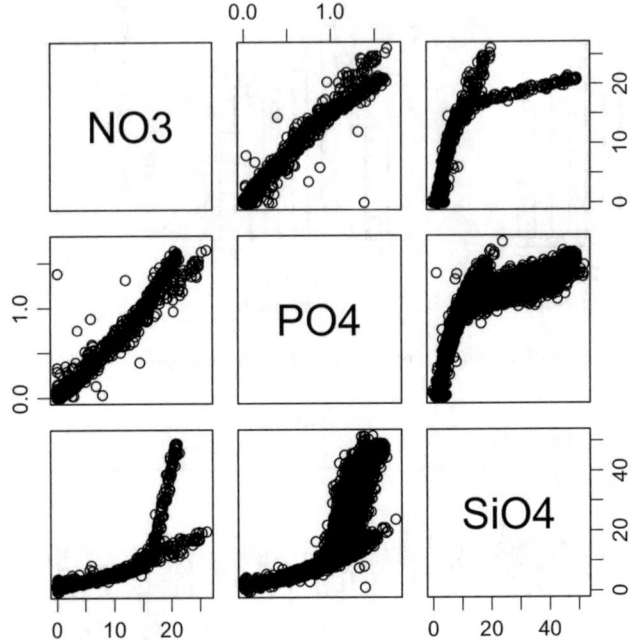

Fig. 2.15 Pairs plot of hydrographic data in the section dataset

2.4.11   Pairs Plots

The R function pairs() makes it easy to "plot everything versus everything else," to quote advice the author received when he entered oceanography. This may be illustrated with the nutrient data in the WOCE oceanographic section provided in the section dataset, e.g.

```
data(section, package="oce")
d <- data.frame(NO3=section[["nitrate"]],
                PO4=section[["phosphate"]],
                SiO4=section[["silicate"]])
```

yields a dataset with NO_3, etc., and then

```
pairs(d)
```

yields Fig. 2.15. Actually, using plot() would have yielded a similar plot, because the first argument is a data frame, and so the generic function would have handed control to pairs().

Exercise 2.28 Use the panels argument to draw the panels as density diagrams, using smoothScatter(). (See page 197 for a solution.)

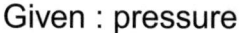

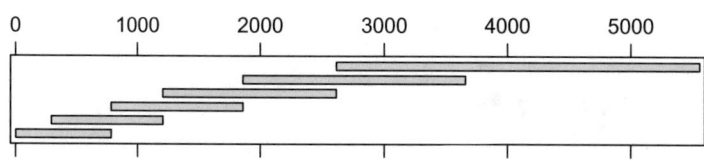

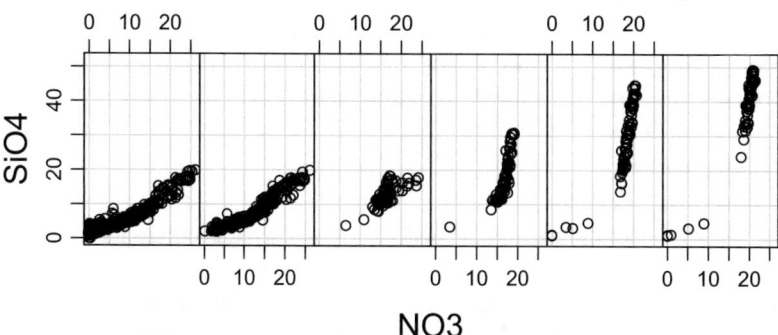

Fig. 2.16 Conditioning plot showing the dependence of silicate on nitrate and depth (indicated by pressure), for the `section` dataset, spanning the Atlantic at 36N. The scatter-plot panels correspond to pressure ranges indicated by bars in the upper panel, revealing that the relationship between SiO$_4$ and NO$_3$ varies with depth in the water column

2.4.12   Conditioning Plots

A conditioning plot, or coplot, represents multivariate dependencies by breaking data into categories of the independent variables. The dependence is expressed with a formula that is similar to that used for regression.

For example, Fig. 2.15 reveals a fairly tight fit between NO$_3$ and PO$_4$, but SiO$_4$ displays a more forked dependency. To investigate whether depth may be the hidden factor here, we might construct a conditioning plot of SiO$_4$ conditioned on both NO$_3$ and pressure. The first step is to extract the data

```
data(section, package="oce")
pressure <- section[["pressure"]]
NO3 <- section[["nitrate"]]
SiO4 <- section[["silicate"]]
```

after which Fig. 2.16 may be constructed with

```
coplot(SiO4 ~ NO3 | pressure, rows=1)
```

providing an indication that the nitrate-silicate relationship indeed varies with depth. Further investigation (e.g. of dependence on latitude) can be handled either by conditioning on another variable or by colour-coding symbols. (Other methods for studying watermass patterns are dealt with in Sect. 5.2.3.)

2.4.13 Function Plots

Given a need to graph the decaying oscillatory function $e^{-x} \cos 2\pi x$, one might start by defining the function

```
f <- function(x) exp(-x) * cos(2 * pi * x)
```

moving on to create vectors to be plotted

```
xx <- seq(0, 1, length.out=100)
yy <- f(xx)
```

and finally plotting

```
plot(xx, yy, type="l")
```

Although this is straightforward, it is certainly more tedious than using a function plot, e.g.

```
plot(function(x) exp(-x) * cos(2 * pi * x))
```

Readers who try this will notice that a default range of x has been used, and that the vertical axis is automatically labelled with the formula. This second feature ensures that the label represents the function faithfully, which reduces the chance of erroneous results in interactive work involving trials with different functions. Small conveniences such as this help to explain why R is widely regarded as a comfortable tool for everyday analysis.

2.4.14 Aesthetic Control of R Graphics

R offers control over many aesthetic aspects of plots, through arguments to the plotting functions and through calls to par(). A few important examples are listed below.

The type argument of plot() indicates whether data on x-y graphs should be indicated with symbols, connected with lines, etc. Symbol type is set with pch and symbol size with cex. Line type and width are set with lty and lwd, respectively. Colour is controlled with col. (For filled symbols, col applies to the perimeter, with bg being used for the fill colour.)

The default R plot margins are wider than those in typical publications, so the margins in most of the examples in this book are narrowed by adjusting the mar argument to par(). The related mgp argument, which controls the spacing of axis titles and labels, is also adjusted here. Readers may find

```
par(mar=c(3, 3, 1, 1), mgp=c(2, 0.7, 0))
```

to be a good starting point, if they wish to devote more space to the content and less to the axes.

Another practical consideration is the need for mathematical notation in plots, which is handled with expressions, e.g.

```
plot(1:10, (1:10)^2, xlab="x", ylab=expression(x^2))
```

An asterisk between symbols in expressions indicates multiplication, so that, for example, `expression(a*b)` is typeset as ab (as in TEX and LATEX). Most other arithmetic operators are typeset as-is, e.g. `expression(a/b)` yields a/b. Indexing gives subscripts, so `expression(a[0])` yields a_0, while carat gives superscripts, `expression(a^2)` yielding a^2. Text and expressions may be combined with `paste()` or multiplication, e.g. both `expression(paste(x, " [m]"))` and `expression(x*" [m]")` yield x [m]. A restriction worth noting is that expressions cannot contain certain keywords, so that, e.g. `expression(V[in])` is disallowed because `in` has a special meaning. However, `substitute()` solves this problem, with, e.g. `as.expression(substitute(V[x], list(x="in")))` yielding V_{in}.

In addition to aesthetic concerns, graphical designer should also take into account issues of accessibility. This is an area that is treated seriously in the R community, and an argument can be made that R outperforms other software in terms of accessibility to viewers with vision limitations (Cleveland and McGill 1984; Ihaka 2003; Zeileis et al. 2009).

2.4.15 Limitations of R Graphics

The first impression of many Matlab users is that R is weak at interactive graphics. This is partly, but not entirely, true. For example, the `zoom` package provides a simple way to zoom in on subregions of plots. Similarly, `locator()` can be used to find plotted values based on mouse clicks. At a deeper level, `getGraphicsEvent()` provides access to mouse up/down events, keyboard events, etc., so it can be used for sophisticated interactive graphs. More widely, and arguably more importantly, the `shiny` package provides a way to set up flexible and powerful GUI systems (see Sect. 2.8). However, all of this requires extra programming, so it is not analogous to the scrolling/zooming behaviour that Matlab users may enjoy.

Matlab users who rely heavily on the GUI may find that they can derive benefits from changing their expectations and adopting new habits. For example, expanding a subregion of a graph in an interactive R session is no harder than using the up-arrow to recall the command that created the plot, adding or modifying the `xlim` or `ylim` arguments to that command, and pressing "enter" to repeat the `plot()` call. This is not as easy as using the mouse to select a region to enlarge, but neither is it especially arduous. Furthermore, R analysts tend to copy their interactive code into scripts, and this means that the details of the graphical setup are easily shared across research groups. This script-based way of working is employed by serious analysts in both Matlab and R, because it has clear advantages in terms of reproducible research. This is a matter of increasing concern in oceanographic research, with journals encouraging authors to supplement their papers with both data and methodological details.

Another complaint from some Matlab users is that multi-panel plots are handled awkwardly in R. Again, this is true, but it does not usually pose serious limitations in practice. Multi-panel graphs may be created in several ways in core R. The most common scheme may be to use par(), specifying mfrow or mfcol to lay out a uniform grid, or using layout() for a non-uniform grid. Unfortunately, these schemes do not permit alteration of previously drawn panels, a task that is accomplished easily in Matlab. The solution in R is simply to plan a bit more, or to cut/paste interactively entered commands into a script, and to make modifications there.

All of the above (and most of this book) is framed in reference to the system called "base graphics." This system is decades old, and has stood the test of time, despite some limitations. An recent alternative is the "grid graphics" system, developed by Paul Murrell, and described in his textbook (Murrell 2006) as well as in the documentation for the grid package. As a low-level system, grid imposes a high burden on the user. The best way to leverage its strengths may be with Hadley Wickham's ggplot2 package, which produces elegant results for a broad range of graphical elements. While ggplot2 is worth serious consideration for many applications, there are two reasons why it is not used in this book or in the companion oce package. First, the R documentation uses base graphics, so serious use of ggplot2 requires that analysts become comfortable with two systems that differ widely in fundamental respects. Second, ggplot2 is significantly slower than base graphics, which can prove to be problematic for the interactive analysis of oceanographic datasets, which are often quite large. For example, on the author's computer, plotting a histogram of 10^7 points took 0.5 s with hist(x) in base graphics, and 7.5 s with qplot(x, geom="histogram") in ggplot2. The slowness of the ggplot2 version is a problem for modern oceanographic instruments, some of which have on-board storage of 16 Gbyte.

2.5   Probability and Statistics

2.5.1   Probability

R is used commonly in undergraduate teaching of probability and statistics, so nobody should not be surprised that it handles these topics elegantly and efficiently. Readers with textbook knowledge and limited coding experience are cautioned against trying to improve upon the R functions, because they tend to be more robust than naive solutions. The choose() function provides a case in point, with

```
choose(52, 5)
 [1] 2598960
```

giving the number of possible 5-card poker hands in a 52-card deck. Mathematically, this is $^{52}C_5 = 52!/(47!\,5!)$, but expressing this directly, with

```
factorial(52) / (factorial(47) * factorial(5))
| [1] 2598960
```

is problematic. Although this works for a 52-card deck, it fails with much larger decks, for which `factorial()` produces values too large to represent in the computer. (Choosing from four decks illustrates this on a 64-bit machine.) Similar limitations exist for other "natural" expressions in probability. The lesson is that only experts should consider replacing built-in R functions.

Exercise 2.29 The Rink Ratz® hockey card game has a 69-card deck with 2 desirable "miraculous save" cards. At the start of the game, 5 cards are discarded without being examined. What is the probability that there will be exactly 1 miraculous save card left in the deck? (See page 197 for a solution.)

2.5.2 Statistics

R provides various properties of statistical distributions, in a family of functions with names that might seem cryptic at first. The first letter of the name indicates the quantity to be calculated, with subsequent letters indicating the distribution, as outlined in Table 2.1. Using d for the first letter retrieves a probability density

Table 2.1 Statistical distributions provided in R

Name code	Distribution
beta	beta distribution
binom	Binomial distribution
cauchy	Cauchy distribution
chisq	χ-squared distribution
exp	Exponential distribution
f	F distribution
gamma	Gamma distribution
geom	Geometric distribution
hyper	Hypergeometric distribution
lnorm	Log-normal distribution
logis	Logistic distribution
nbinom	Negative-binomial distribution
norm	Normal distribution
pois	Poisson distribution
signrank	Signed-rank distribution
t	t distribution
unif	Uniform distribution
weibull	Weibull distribution
wilcox	Wilcox distribution

function, while p retrieves a cumulative probability density function, q a quantile function, and r a random number function. The examples of the following sections should clarify the scheme.

2.5.2.1 Statistical Tables

Although statistical tables lose much of their value in a system that permits easy calculation of any desired quantity, readers might find it instructive to try reproducing some of the tables found in their textbooks, e.g.

```
t(outer(1:3, 1:5, function(nu1, nu2) qf(1-0.05, nu1, nu2)))
            [,1]         [,2]         [,3]
[1,] 161.447639 199.500000 215.707345
[2,]  18.512821  19.000000  19.164292
[3,]  10.127964   9.552094   9.276628
[4,]   7.708647   6.944272   6.591382
[5,]   6.607891   5.786135   5.409451
```

creates a table of F values for $\alpha = 0.05$. Here, outer() has been used to create the table, with an anonymous function taking v_1 and v_2 as input.

Exercise 2.30 Construct a graph comparing the normal distribution with the t distribution with 2 degrees of freedom. (See page 197 for a solution.)

2.5.2.2 Confidence Intervals of Means

A confidence interval on the mean of a vector x of length n drawn from the Student t distribution is given by $t_* s / \sqrt{n}$ where t_* is the value of the t distribution for $n - 1$ degrees of freedom and a specified confidence level, and the standard deviation s may be computed with sd(). For example,

```
qt(0.975, df=length(x)-1) * sd(x) / sqrt(length(x))
```

illustrates the computation of a 95% confidence interval of the mean of x.

Other distributions are handled similarly, with qt() being replaced with qnorm() for the normal distribution, dchisq() for χ^2, etc.

2.5.2.3 Measurement Uncertainty and Error Bars

It is important to distinguish between the confidence intervals of means discussed in the previous section, and measurement uncertainties. The latter refer to the spread of data, and this is usually what is meant by with "±" in text and by error bars on plots.

As explained by Taylor and Kuyatt (1994, Sections 3 and 4), measurement uncertainties may be divided into Type A, related to the statistics of the data, and Type B, related to wider factors such as previous measurements, manufacturer specifications, calibrations, etc.

For Type A, Taylor and Kuyatt (1994, Section 2.3) recommend calculating measurement uncertainty as ku_i, where k is a "coverage factor" calculated from the statistics of the data and u_i is the "standard uncertainty," inferred as the standard deviation and thus calculated with sd(). This is a factor of $\sqrt{n}$ larger than the confidence interval on the mean, so analysts must be careful to report methods clearly to avoid confusion between two numbers that may be very different.[16]

Setting k to 1.96 achieves 95% coverage probability for normally distributed data, perhaps explaining why some analysts use $k = 2$ for approximate 95% coverage probability. Different applications may call for different k values, so analysts should provide enough information to let readers perform their own calculations with different parameters (Taylor and Kuyatt 1994, Section 7).

Exercise 2.31 Write a function that computes measurement uncertainties assuming a t distribution. (See page 198 for a solution.)

Exercise 2.32 Write a function that plots error bars. (See page 198 for a solution.)

2.5.2.4 Random Number Generation

As noted above, random numbers may be generated from many different distributions. Two commonly used cases are rnorm() for the normal distribution and runif() for the uniform distribution. Another useful function is set.seed(), which sets the seed for a sequence of random numbers, which is handy for reproducible research.

2.5.3 Summaries and Overview Functions

The summary() function produces useful summaries of data and objects. Being a generic function, it acts differently when supplied with different arguments. In the simple case of numerical arguments, it reports the data range, mean, and three quartiles, e.g.

```
summary(rnorm(100))
    Min.  1st Qu.   Median     Mean  3rd Qu.     Max.
-2.48865 -0.51127 -0.02768 -0.04943  0.60007  2.55300
```

---

[16]Home electricity provides a dramatic illustration. Although voltage measurements may give a confidence interval on the mean that barely departs from 0V, the measurement uncertainty will indicate that any given measurement could easily be of order 100V. That is why electrical outlets must be covered up, in houses with young children.

Other data types are treated in sensible ways. Related functions are also useful, e.g. quantile() calculates quantiles, and fivenum() gives Tukey's five-number summary (lower limit, lower "hinge", median, upper hinge, and upper limit). A useful semi-graphical way to summarize data is with stem(), e.g. with[17]

```
set.seed(253)
x <- rnorm(n=30, mean=2, sd=0.1)
```

as data, the results

```
stem(x)

  The decimal point is 1 digit(s) to the left of the |

  18 | 9
  19 | 13344
  19 | 56778888
  20 | 12333
  20 | 666779
  21 | 111
  21 |
  22 | 1
  22 |
  23 | 3
```

indicate (in the first and second lines) that 1.89 and 1.91 are in x, along with two instances of 1.93 and 1.94, etc. Thus, stem() produces not just a textual histogram, but also a list of the (rounded) data.

There are several graphical representations that may be used, e.g. Fig. 2.17 shows histograms and box plots

```
hist(x, main="")
boxplot(x, horizontal=TRUE, notch=TRUE)
```

along with density and violin plots

```
plot(density(x), main="")
rug(x, side=1)
library(vioplot)
vioplot(x, horizontal=TRUE, col="gray")
```

Note that the data are shown along the lower axis of the density plot, by using rug(). This function is quite handy with relatively small datasets.

2.5.4 Hypothesis Testing

Hypothesis testing is supported with a suite of R functions. Explanations of the underlying ideas may be found in most textbooks on statistics, and e.g. Legendre and

---

[17]Note the use of set.seed() to let readers reconstruct the example.

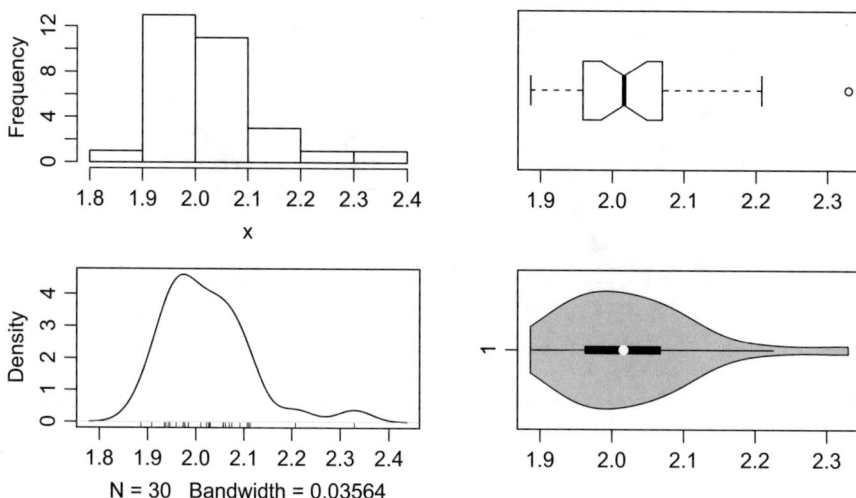

Fig. 2.17 Data summaries from hist(), boxplot(), density() and vioplot()

Legendre (1998) and Borcard et al. (2011) provide treatments in the oceanographic context.[18]

The popular *t* test is handled with t.test(), which provides both one-sample or two-sample tests, the latter paired or unpaired. The alternative argument controls alternative hypotheses, while mu sets the mean, conf.level sets the confidence level, etc.

The data shown in Fig. 2.17 provide an example.

```
t.test(x)
            One Sample t-test

data:  x
t = 120.05, df = 29, p-value < 2.2e-16
alternative hypothesis: true mean is not equal to 0
95 percent confidence interval:
 1.989308 2.058263
sample estimates:
mean of x
 2.023786
```

---

[18]It is unwise to use hypothesis tests without considering their limitations. Some issues of misapplication are outlined by, e.g., Johnson and Omland (2004) and Hauer (2004), and deep concerns about the misuse of *p* values are raised in a highly influential editorial in The American Statistician (Wasserstein and Lazar 2016).

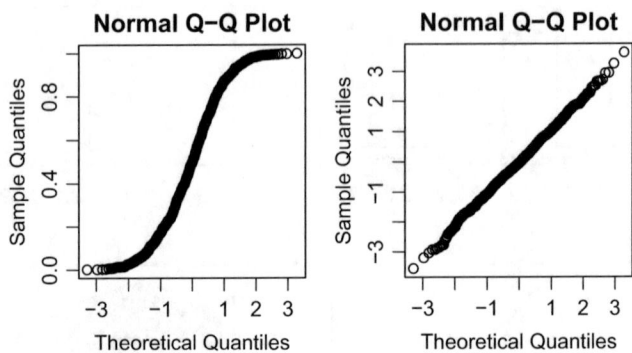

Fig. 2.18 Q-Q plots for random numbers chosen from a uniform distribution (left) and normal distribution (right)

The *p* value is small[19] compared with typical significance levels (often 0.05), which argues in favour of the alternative hypothesis (which is helpfully stated in the t.test() output) that the mean of *x* is not equal to zero.

The documentation for the stats package is worth consulting for its discussion of the many other tests that complement the *t* test. Just one more example will be given here: testing for normality.

A typical first step in testing a distribution for normality is to draw a Q-Q plot with qqplot(). For example, Fig. 2.18 shows such plots with random numbers drawn from uniform and normal distributions.

```
set.seed(254)
uniform <- runif(n=1e3)
normal <- rnorm(n=1e3)
par(mfcol=c(1,2))
qqnorm(uniform)
qqnorm(normal)
```

A linear Q-Q plot suggests a normal distribution, as in the right panel. This plot style is so useful that it is included in the graphs produced when plot() is called with the results of a regression model (see Sect. 2.5.5).

R provides a variety of formal tests as well, e.g. a Shapiro-Wilks test

```
shapiro.test(uniform)
            Shapiro-Wilk normality test

data:  uniform
W = 0.9495, p-value < 2.2e-16
```

yields a small *p* value, suggesting (as expected) that the uniform data are not normally distributed. By contrast, the same test on the normal data yields a high *p*

---

[19]If $p < 2.2 \times 10^{-16}$, R regression summaries simply reports "p-value: < 2.2e-16".

value, consistent with a normal distribution. (Other tests of normality are provided in the `mvShapiroTest`, `nortest` and `fBasics` packages.)

Exercise 2.33 Show how `split()` and `laply()` can be used to produce a monthly climatology of a signal, and illustrate using the results of Exercises 2.31 and 2.32. (See page 198 for a solution.)

2.5.5 Regression

Since regression is used in many fields, there are many resources that explain the theory and provide practical advice. Accordingly, the present treatment is somewhat cursory, focussing mainly on R details; see also Chambers and Hastie (1992), Faraway (2005), and Faraway (2002).

The standard R system provides linear, generalized, and nonlinear least-squares regression, with `lm()`, `glm()`, and `nls()` respectively, Optional packages provide much more, e.g. MASS (Venables and Ripley 1999) provides robust regression with `rlm()` and ridge regression with `lm.ridge()`, `segmented` (Muggeo 2008) provides piecewise-linear regression, etc. Some methods are handled by multiple packages, e.g. ridge regression is provided by `mgcv` (Wood 2001) as well as by MASS. Readers with specific needs should study the documentation of specialized functions, but it is wise to start with the basics, the following discussion of which is divided into linear and nonlinear categories.

2.5.5.1 Linear Regression

Linear regression involves the study of data modelled by

$$y = \beta_1 x_1 + \beta_2 x_2 + \cdots + \epsilon \qquad (2.3)$$

where y is a response vector that depends on vectors x_i, scalars β_i are constants to be determined, and ϵ is a vector of error or misfit. An intercept is handled by setting $x_1 = 1$, and polynomial regression by expressing, e.g., $y = a + bx + cx^2$ with $x_1 = 1$, $x_2 = x$ and $x_3 = x^2$.

Synthetic data can be used to illustrate how regression (essentially, calculating β_i) works in R. For example,

```
set.seed(2551)                          # for reproducibility
x <- seq(0, 1, length.out=25)
y <- 1 + 2 * x + 4 * x^2 + rnorm(length(x), sd=0.1)
```

creates a test case that is not quite linear, so it can be used to illustrate how to work through a sequence of regression models.

A sensible first step is to plot the data, e.g.

```
plot(x, y)
```

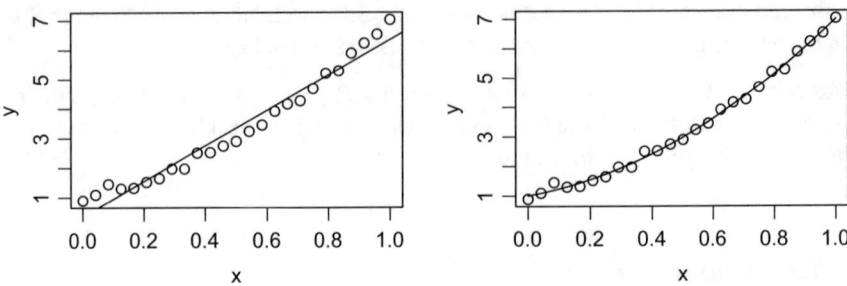

Fig. 2.19 Use of `lm()` for linear and quadratic regression models

producing the symbols in Fig. 2.19. Although inspection reveals curvature, an analyst working with such data might still start with a linear model, in the interests of simplicity or to satisfy a theoretical or practical constraint.

The workhorse regression function for linear models is `lm()`, and

```
lm(y ~ x)
Call:
lm(formula = y ~ x)

Coefficients:
(Intercept)                    x
     0.3469               5.9952
```

shows that it is easy to use. The "`~`" in the first argument indicates that it is a formula. It could be read as "y depends linearly on x", but emphatically not that the model is $y = \beta x$, because an intercept is implied; the zero-intercept case is explained later. A formula can specify interaction terms and other complications; see `help(formula)` and `help(lm)` for more on the notation.

The results of the regression may be saved to a variable, e.g.

```
linear <- lm(y ~ x)
```

and this is helpful because it facilitates further analysis, e.g.

```
abline(linear)
```

adds the regression line to the left panel of Fig. 2.19.

The right panel of Fig. 2.19 shows the result of a quadratic fit, resulting from using `poly()` in the regression formula

```
quadratic <- lm(y ~ poly(x, 2, raw=TRUE))
```

Here, `raw=TRUE` tells `poly()` to use conventional polynomials, not orthogonal polynomials. Another way to get conventional polynomials is with

```
quadratic <- lm(y ~ x + I(x^2))
```

where `I()` causes `lm()` to take "`^`" to mean exponentiation as opposed to interaction (see `help(formula)`). The prediction may be added with

```
plot(x, y)
lines(x, predict(quadratic))
```

where it should be noted that `predict()` is a generic function that works for all sorts of regressions.

The `summary()` function is useful for studying regression results, e.g.

```
summary(linear)
Call:
lm(formula = y ~ x)

Residuals:
    Min      1Q  Median      3Q     Max
-0.4512 -0.3282 -0.0916  0.2943  0.6782

Coefficients:
            Estimate Std. Error t value Pr(>|t|)
(Intercept)    0.347      0.140    2.47    0.021 *
x              5.995      0.240   24.94   <2e-16 ***
---
Signif. codes:
0 '***' 0.001 '**' 0.01 '*' 0.05 '.' 0.1 ' ' 1

Residual standard error: 0.361 on 23 degrees of freedom
Multiple R-squared:  0.964,        Adjusted R-squared:  0.963
F-statistic:  622 on 1 and 23 DF,  p-value: <2e-16
```

provides the information normally stated in publications, and much more. First, it repeats the regression formula, which is helpful for stored output. Then, it provides information about the residual deviations from the fit, which can be quite helpful if combined with information on measurement uncertainties. The model coefficients are presented next, along with standard errors and corresponding t and p values. Finally, the overall fit is described in terms of correlation coefficients, F statistic and p value.

In this case, the artificial data have been constructed with a random y component of standard deviation 0.1, and the residual standard error of the linear regression is three times as large. Such a comparison (with the standard deviation of this simulation perhaps being replaced with an estimate of measurement uncertainty in a real study) might motivate further investigation of model formulation, even if Fig. 2.19 had not shown a systematic misfit.

The improved results of quadratic regression are revealed by

```
summary(quadratic)
```

an excerpt of which is

```
Residual standard error: 0.08655 on 47 degrees of freedom
```

indicating a residual comparable to the "noise," a condition that might indicate success if this were a calibration study.

However, `summary()` is not the only tool R provides for checking on regression results. The `plot()` function is specialized to handle regression output, e.g.

```
par(mfrow=c(2, 2))
plot(linear)
```

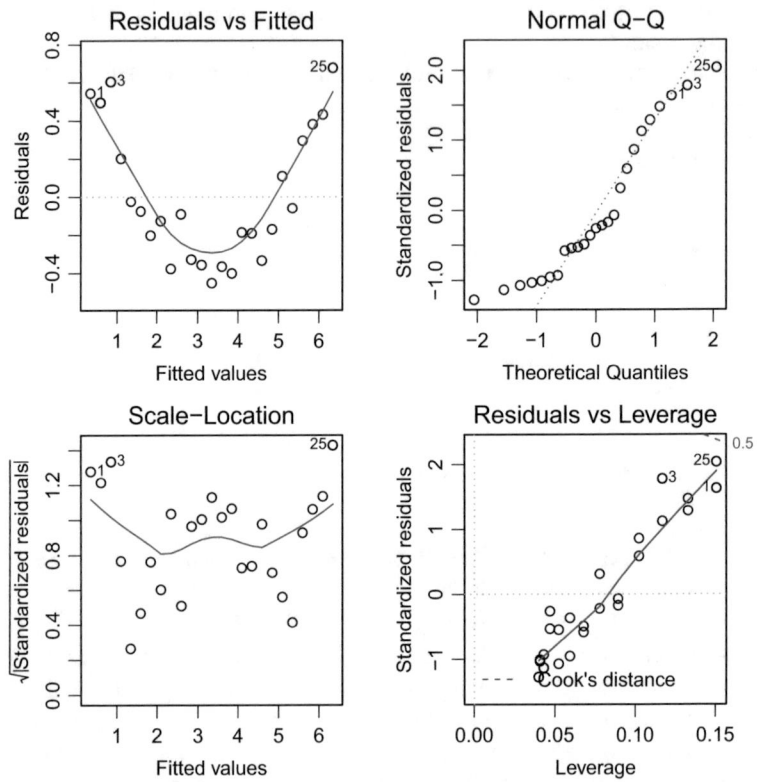

Fig. 2.20 Plot of the regression object for the linear fit shown in Fig. 2.19

yields Fig. 2.20. The top-left panel shows how the model-data misfit varies with fitted value (a format that also handles multiple independent variables), and the clear pattern of variation indicates the limitation of a linear model in this case. Furthermore, the Q-Q plot in the top-right panel reveals that the misfit is not normally distributed. The other two panels provide further diagnostics (see help(plot.lm)). The numbers in the panels are the indices of points that may deserve further consideration, and this can be helpful in revealing data outliers, as well as poor choices of regression model.

A few questions should be answered before discussing the nonlinear case.

- *Can the line be forced to go through the origin?* Yes, by appending -1 to the formula, e.g. writing lm(y ~ x - 1) in place of lm(y ~ x).
- *Can the line slope be specified?* Yes, by using offset. This can be useful in fitting power laws, e.g. for $y = Ax^{2/3}$ with $A = 10$ plus noise,

```
x <- 1:100
y <- rnorm(100, mean=10)*x^(2/3)
```

```
m <- lm(log10(y) ~ offset(log10(x^(2/3))))
10^confint(m)
              2.5 %    97.5 %
(Intercept) 9.757443 10.20824
```

where `confint()` gives confidence intervals on the regression coefficients.
Another notation places known dependence on the left in the formula, e.g.

```
10^confint(lm(log10(y/x^(2/3)) ~ 1))
```

- *Can R do stepwise regression?* Yes; see `help(step)` and `help(stepAIC, package="MASS")`.
- *Can R draw confidence intervals on the fitted curve?* Yes, with `predict()`

```
prediction <- predict(lm(y~x),interval="confidence")
lines(x, prediction[,1]) # fit
lines(x, prediction[,2], lty="dashed") # lower bound
lines(x, prediction[,3], lty="dashed") # upper bound
```

- *Can R do multiple regression?* Yes, just add the independent variables to the right side of the formula, e.g.

```
data(ctd, package="oce")
sigthe <- ctd[["sigmaTheta"]]
sal <- ctd[["salinity"]]
temp <- ctd[["temperature"]]
m <- lm(sigthe ~ sal + temp)
```

- *How are regression coefficients retrieved?* With m as in above, use, e.g.

```
coef(m)
(Intercept)         sal          temp
 -6.6899404    1.0215729    -0.1170753
```

- *Can R handle type-II regression, with errors in both x and y?* Yes, and several methods are available. Several R packages support type-II regression, including `smatr` (Warton et al. 2012) and `lmodel2` (Legendre 2014). See Warton et al. (2006) for general issues, and e.g. Ricker (1973), Marsden (1999), McArdle (2003), and Clarke and Van Gorder (2012) in the context of oceanography.
- *Can R handle robust regression?* Yes, using `rlm()` from the MASS package, e.g. the following simulates the effect of a missing-value code

```
x <- 1:10
y <- -10 + 5 * x + rnorm(10)
y[10] <- -99.99 # insert 'missing value' code
```

and Fig. 2.21, constructed with

```
plot(x, y)
abline(lm(y ~ x))
library(MASS)
abline(rlm(y ~ x), lwd=3)
```

shows that `rlm()` performs better than `lm()` in this case.
- *Can R handle piecewise linear regression?* Yes, with the `segmented` package, as the right panel of Fig. 2.21 shows for wave heights and wind speeds

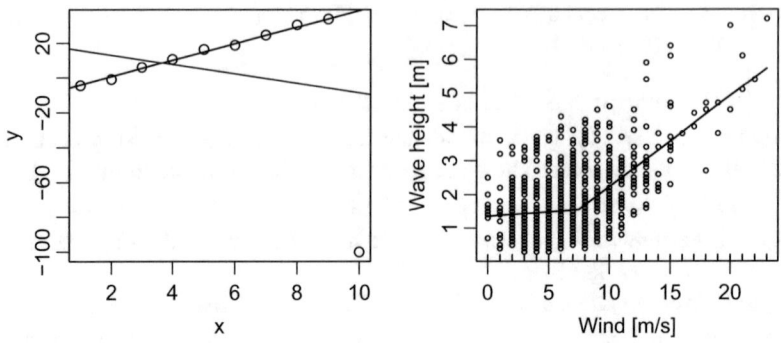

Fig. 2.21 Left: conventional and robust regression of data with a spurious outlier, in thin and thick lines. Right: segmented regression of buoy data

```
data(buoy, package="ocedata")
plot(buoy$wind, buoy$height, cex=1/2,
     xlab="Wind [m/s]", ylab="Wave height [m]")
library(segmented)
s <- segmented(lm(height~wind, data=buoy),
               seg.Z=~wind, psi=c(10))
plot(s, add=TRUE, lwd=3)
```

Exercise 2.34 Contrast the residual plots produced by plot() for linear and quadratic. (See page 199 for a solution.)

Exercise 2.35 Use eigen() and cov() to draw a line that intersects the means of x and y, and that has the same slope as the principal eigenvector of the covariance matrix. (See page 200 for a solution.)

2.5.5.2   Nonlinear Regression

Nonlinear regression involves working with a model equation of the form

$$y = f(\mathbf{x}, \boldsymbol{\beta}) + \epsilon \qquad (2.4)$$

where f is a nonlinear function of independent variables $\mathbf{x} = (x_1, x_2, \dots)$ and model parameters $\boldsymbol{\beta} = (\beta_0, \beta_1, \dots)$ and ϵ represents model-data misfit. The concepts of nonlinear regression are explained in a variety of textbooks and other sources; in the R context, see, e.g., Chapter 8 of Venables and Ripley (1999) and the references in the documentation for nls(), the function used for nonlinear regression in R. Before getting into the details, it should be noted that nls() is harder to use than its linear cousin lm(), simply because nonlinear regression is a much more complicated task, and not one that can always be accomplished in practice (see, e.g., Gallant 1975).

Fig. 2.22 An oxygen concentration profile and the predictions of a nonlinear regression model

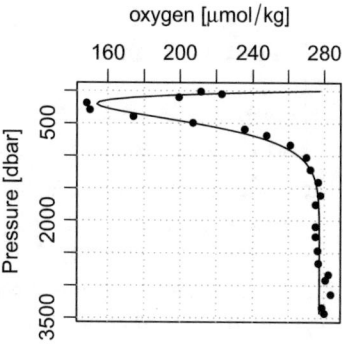

Typical procedures may be illustrated with the sample task[20] of fitting a model to an oxygen profile plotted in Fig. 2.22 with

```
data(section, package="oce")
stn <- section[["station", 112]]
plotProfile(stn, "oxygen", type="p", pch=20)
p <- stn[["pressure"]]
O2 <- stn[["oxygen"]]
```

This suggests a model function with a mid-depth minimum, e.g.

$$O_2 = A - Bpe^{-p/C} \tag{2.5}$$

where A, B and C are parameters to be determined. Since nls() requires starting values for its parameter search, the first step is to estimate reasonably sensible values of A, B and C (e.g. all should be positive, to get $O_2 > 0$ with a mid-depth minimum and a deep-water asymptote). Substituting $p = 0$ in (2.5) reveals that A is the surface O_2 value, and differentiation indicates that the minimum $O_2 = A - BC/e$ occurs at $p = C$. Figure 2.22 might thus motivate starting values $A = 300\,\mu$mol/kg, $B = 2\,\mu$mol/(kg dbar), and $C = 200$ dbar, each perhaps to within a factor of 2. A test of whether nls() can find a solution without accurate starting values might be to set each of the three numerical values to 10.

Indeed, these starting values produce a converged solution

```
m <- nls(O2~A-B*p*exp(-p/C), start=list(A=10, B=10, C=10))
```

and the gist of this nonlinear regression is summarized as

```
m
Nonlinear regression model
  model: O2 ~ A - B * p * exp(-p/C)
   data: parent.frame()
       A        B        C
```

---

[20]See also the NISTnls package, which provides data and code for statistical test suites developed by researchers at the U.S. National Institute for Standards and Technology.

```
277.473    1.668 201.307
  residual sum-of-squares: 3501

Number of iterations to convergence: 11
Achieved convergence tolerance: 3.894e-06
```

The model prediction is added to Fig. 2.22 with

```
pp <- seq(0, max(p), length.out=100)
lines(predict(m, list(p=pp)), pp)
```

and the reasonable agreement with data (compared with O_2 scatter in deep water) might motivate further investigation. The next step would be to use summary(m), after which using profile(m) and plot(profile(m)) can shed light on the tightness of the fitted parameters. However, before taking any further steps, nls() must converge, which is not always the case. Problems fall into several classes.

The first requirement is that start must match the formula. The regression cannot succeed if start is missing parameters, with, e.g.

```
nls(y ~ a+b*x, start=list(a=1))
```

yielding the straightforward message

```
Error in nls(y ~ a + b * x, start = list(a = 1)) :
   parameters without starting value in 'data': b
```

Unfortunately, specifying too many start items

```
nls(y ~ a + b * x, start=list(a=1, b=1, c=1))
```

yields the more cryptic error message

```
Error in nlsModel(formula, mf, start, wts) :
   singular gradient matrix at initial parameter
   estimates
```

This is because the calculation involves partial derivatives of model misfit with respect to each start item, and the derivative with respect to c is zero for this model, yielding a non-invertible matrix.

It is also a mistake for models to have commingled parameters, e.g. the $y = (\beta_1 + \beta_2)x$ cannot be solved for β_1 and β_2 independently, so

```
x <- 1:10
y <- 3*x
nls(y ~ (a+b)*x, start=list(a=1, b=2))
```

yields a singular-gradient matrix. In this case, the matrix is singular because two derivatives are identical, and matrices with identical rows or columns cannot be inverted.

Helpfully, nls() also checks for singularity during its search through parameter space. For example, $y = ax + \exp(bx)$ degenerates to the problematic form $y = (a + b)x$ if nls() selects a value of b for which $|bx| \ll 1$ for the data being examined. The error message in such a condition is similar to that shown above, but without the "initial" phrase.

Problems can also arise when the data-model misfit function has a much stronger dependence on one parameter than on another. This is akin to the challenge of navigating to the lowest spot in a curvy valley that is long and thin. This sort of problem can be signalled by a variety of error messages from nls(), e.g.

```
number of iterations exceeded maximum ...
```

(where ... will be an integer) indicates that an excessive number of steps has been taken, with no end in sight. This might be solved with, e.g.

```
nls(..., control=list(maxiter=200)) # default is 50
```

but it may be better to reformulate the problem, e.g. by a change of variable to get $O(1)$ variations. Similarly, the problem

```
step factor ... reduced below 'minFactor' of ...
```

may be alleviated by altering minFactor in control.

Sometimes, it helps to try alternative solution methods. The default algorithm employed by nls() is based on a Gauss-Newton procedure, but if this fails, it might help to use the algorithm argument to try other methods. The "port" algorithm is notable because it permits the specification of upper and lower limits for the parameters, which can help if nls() is straying into regions of parameter space that are unphysical (e.g. negative salinities) or uninteresting (e.g. angles outside the range 0 to 2π).

If the nls() numerical differentiation is problematic, a function can be provided to calculate derivatives (Chambers and Hastie 1992), e.g. for (2.5)

$$\frac{\partial O_2}{\partial A} = 1, \quad \frac{\partial O_2}{\partial B} = -p\,e^{-p/c}, \quad \text{and} \quad \frac{\partial O_2}{\partial C} = -\frac{Bp^2}{C^2}e^{-p/C} \tag{2.6}$$

which may be employed as follows:

```
O2Model <- function(A, B, C)
{
    E <- exp(-p / C)
    prediction <- A - B * p * E
    gA <- 1
    gB <- -p * E
    gC <- -B * p^2 / C^2 * E
    gradient <- cbind(gA, gB, gC)
    attr(prediction, "gradient") <- gradient
    prediction
}
mg <- nls(O2~O2Model(A, B, C), start=list(A=10, B=10, C=10))
```

with results as already shown.

As a general matter, it can be helpful to call nls() with trace=TRUE, which prints the sequence of parameter values being tested. This can reveal a variety of problems, e.g. poor starting values can be signalled by rapid departures from the initial state. When a sequence of datasets are to be studied, it can be good to use try() to handle errors in nls() calls, whether to provide helpful information on the problematic test cases or to work through a sequence of trial model formulations.

Exercise 2.36 Extract tritium Tu and pressure p from the `ocedata` dataset `geosecs235`, and use `nls()` to fit the model

$$\text{Tu} = A\exp(-(p - p_0)^2/D^2) + A\exp(-(p + p_0)^2/D^2)$$

where A, p_0 and D are parameters to be inferred. (See page 201 for a solution.)

2.5.6   Analysis of Variance

The analysis of variance (ANOVA) is popular in some branches of oceanography, and barely used in others. Introductions to the method may be found in most Statistics textbooks (see, e.g., De Veaux et al. 2006). The R perspective is described briefly in the documentation of `aov()` and `anova()`, with much more detailed treatments of ANOVA and the related method of regression being provided in the texts by Chambers and Hastie (1992) and Faraway (2005), a version of the latter being freely available online as Faraway (2002). These are all general treatments. For marine examples, see texts by Legendre and Legendre (1998) or Borcard et al. (2011), the second of which uses R.

An example with constructed data is sufficient to reveal procedures. Suppose thermometers labelled $T1$, $T2$ and $T3$ each record temperatures in five isothermal water baths. If the instruments all have random errors of $0.001°C$, $T2$ has a systematic offset of $0.010°C$ and $T3$ has a systematic offset of $0.001°C$, then an artificial dataset can be created with

```
set.seed(256)
T <- data.frame(T1=10.000 + rnorm(n=5, sd=0.001),
                T2=10.010 + rnorm(n=5, sd=0.001),
                T3=10.001 + rnorm(n=5, sd=0.001))
```

A box plot is a good way to display the data, and

```
boxplot(T, horizontal=TRUE, xlab="Temperature")
```

creates the left panel of Fig. 2.23, which suggests that $T2$ is definitely different from the others, and that $T3$ is possibly offset from $T1$.

The work is simplified by using `stack()`, which returns columns for the dependent variable[21] in `values` and the dependent variable `ind`.

```
Ts <- stack(T)
```

The analysis of variance is carried out with `aov()`, using a formula notation like that used by `lm()`

```
a <- aov(values ~ ind, data=Ts)
```

---

[21] An alternative to `stack()` is `melt()`, from the `reshape2` package (Wickham 2007). If this is used, then `aov()` must use `value` for `values` and `variable` for `ind`.

95% family-wise confidence level

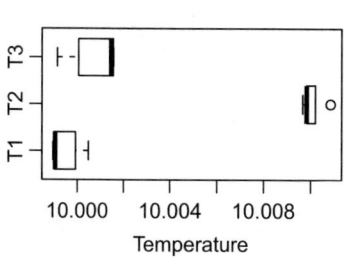

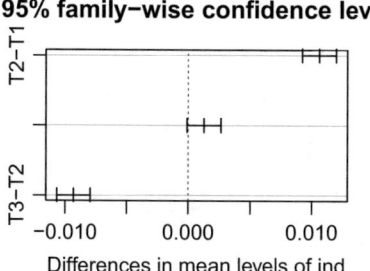

Fig. 2.23 Results of analysis of variance of simulated data from three thermistors

As with linear regression, summary() gives an overview:

```
summary(a)
              Df      Sum Sq    Mean Sq  F value    Pr(>F)
ind            2  0.0003346  1.673e-04    259.3  1.34e-10 ***
Residuals     12  0.0000077  6.500e-07
---
Signif. codes:
0 '***' 0.001 '**' 0.01 '*' 0.05 '.' 0.1 ' ' 1
```

which suggests differences between the instruments. This is consistent with the boxplot of Fig. 2.23, but there is another way to see this graphically, with the Tukey honest significant difference, i.e.

```
plot(TukeyHSD(a))
```

producing the right panel of Fig. 2.23, which suggests that instruments $T1$ and $T3$ are similar, but that $T2$ is different. The procedure has thus flagged the $T2$ anomaly in the constructed data, but not the smaller $T3$ anomaly.

Exercise 2.37 Increase the value of n until the TukeyHSD diagram indicates that $T1$ and $T3$ are producing different values. (See page 201 for a solution.)

2.5.7 Partitioning Decision Trees

The term "regime shift" has been used to describe variations in physical or biological properties of the ocean that take the form of rapid transitions from one quasi-steady state to another. This is a topic of great interest and some controversy; see, e.g., Miller et al. (1994), Rudnick and Davis (2003), deYoung et al. (2008) and Lindegren et al. (2012).

One way to look for regime shifts is with the partitioning tree approach used by ctree() in the party package (Hothorn et al. 2006). Artificial data

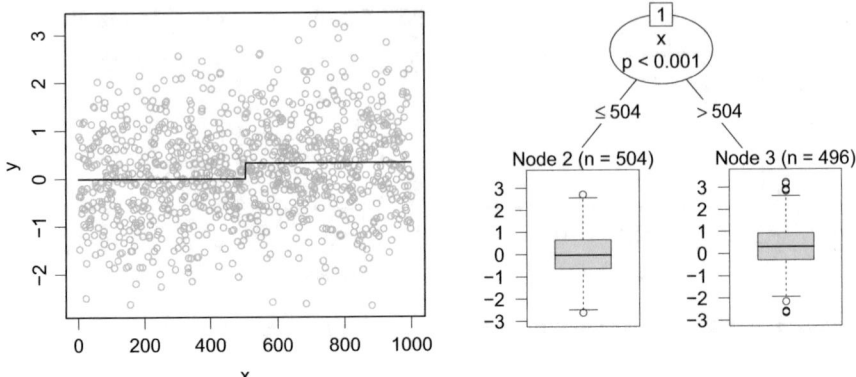

Fig. 2.24 Analysis of a simulated regime shift. Top: random numbers that shift slightly at $x =$ 500. Bottom: diagram produced by `ctree()` in the `party` package

```
set.seed(257)                          # for reproducibility
n <- 1000
x <- 1:n
y <- rnorm(n) + ifelse(x > 500, 1/3, 0)
plot(x, y, cex=0.75, col="gray")
```

as shown in the symbols of the left panel of Fig. 2.24 can illustrate procedures. It is difficult to discern the shift at $x = 500$, but

```
library(party)
p <- ctree(y ~ x)
lines(x, predict(p))
```

adds lines to the diagram that show that `ctree()` finds a shift at $x = 504$, reasonably close to the actual value. More about the fit is obtained with

```
plot(p)
```

which produces the right panel of Fig. 2.24.

Another approach is to use the `changepoint` package (Killick and Eckley 2014; Killick et al. 2016), e.g. a test for a shift in the mean

```
library(changepoint)
cpts(cpt.mean(y, penalty="SIC"))
[1]  504
```

matches the `ctree()` result.

Exercise 2.38 Use `ctree()` and `cpt.mean()` to examine the Southern Oscillation Index data (`soi` in the `oce` package) for regime shifts between 1967 and 1985. (See page 202 for a solution.)

2.6 Numerical Methods

2.6.1 Sorting

The sort() function returns a sorted vector, while order() returns indices that will yield a sorted vector. Thus, for example, x[order(x)] yields the same results as sort(x). The order() method has the advantage of working with matrices, e.g. the ocean data

```
data(oceans, package="ocedata")
```
may be reordered according to average depth with

```
oceansOrdered <- oceans[order(oceans$AvgDepth), ]
```
and a reverse ordering can be found by supplying decreasing=TRUE to order(). Ranking can be achieved by a nested call to order(), e.g. a column of rank by average depth may be added with

```
oceans$rankByAvgDepth <- order(order(oceans$AvgDepth,
    decreasing=TRUE))
```

2.6.2 Root Finding

As noted in Sect. 2.3.11.4, roots of univariate functions may be found with uniroot(). Roots of polynomials can be found with polyroot(), e.g. $(x - 1)(x + 1)$ may be written $a_1 + a_2 x + a_3 x^2$ with $a = (-1, 0, 1)$, yielding:

```
polyroot(c(-1, 0, 1))
 [1]   1+0i -1+0i
```

2.6.3 Integration

Numerical integration of a function of a single variable is handled with integrate(). For example, $\int_0^\pi \sin\theta\, d\theta$ is calculated (along with an error estimate) with

```
integrate(sin, 0, pi)
 2 with absolute error < 2.2e-14
```
Infinite limits may also be supplied, e.g. for the witch of Agnesi function

```
woa <- function(x, a=1)
    8 * a^3 / (x^2 + 4*a^2)
integrate(woa, -Inf, Inf)
 12.56637 with absolute error < 1.3e-09
```
the integral matches the theoretical value of 4π to within 2×10^{-15}.

Exercise 2.39 Use `integrate()` to calculate the perimeter of an ellipse of major axis $a = 2$ and minor axis $b = 1$. (See page 202 for a solution.)

2.6.4   Piecewise Linear Interpolation

Piecewise-linear interpolation is provided with `approx()`, which returns interpolated values, and `approxfun()`, which returns an interpolating function.

A common use of `approx()` is to interpolate values to a uniform one-dimensional grid. For example, the `ctd` dataset in the `oce` package holds hydrographic data sampled at unevenly spaced pressures.

```
data(ctd, package="oce")
p <- ctd[["pressure"]]
```

so that, e.g., salinity may be interpolated to pressures $(0, 0.5, \ldots)$ dbar with

```
S <- ctd[["salinity"]]
Sinterp <- approx(p, S, seq(0, max(p), 0.5))$y
```

The first two arguments to `approx()` are the independent and dependent variables, and the third is the interpolating grid. Values beyond the data range will be returned as `NA`, although the `rule` argument to `approx()` provides an alternative to that convention.

The use of `approxfun()` may be illustrated with the turbulence measurements of Grant et al. (1962). The dataset `turbulence` in the `ocedata` package contains wavenumber k and one-dimensional spectrum function ϕ, provided in non-SI units for comparison with this classic paper. A quantity of interest is ϵ, the rate of viscous dissipation of turbulent kinetic energy per unit mass. Under certain conditions, this is $\epsilon = 15\nu \int_0^\infty k^2\phi \, dk$ where ν is the kinematic viscosity of seawater, motivating a plot of $k^2\phi$ versus k, with

```
data(turbulence, package="ocedata")
k <- turbulence$k
phi <- turbulence$phi
plot(k, k^2*phi, pch=20, ylim=c(0, 0.41),
     xlab=expression(k), ylab=expression(k^2*phi))
```

resulting in Fig. 2.25. Grant et al. (1962) reported that $\epsilon = 0.610 \, \text{cm}^2/\text{s}^3$, and it may be of interest to see whether a similar value can be recovered by integrating under a function representing the data just plotted, e.g.

```
lfcn <- approxfun(k, k^2 * phi)
```

the results of which are added to the diagram with

```
kk <- seq(min(k), max(k), length.out=100)
lines(kk, lfcn(kk))
```

The integration is performed with

```
I <- integrate(lfcn, min(k), max(k))
```

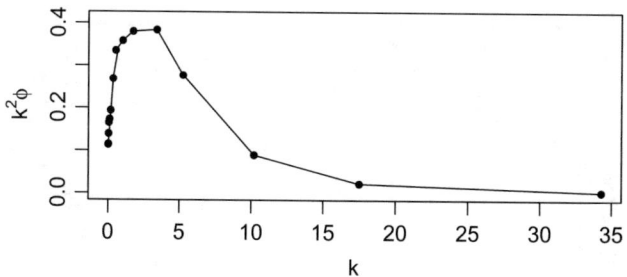

Fig. 2.25 Turbulence data recorded by Grant et al. (1962), with k wavenumber (1/cm) and ϕ one-dimensional velocity spectral function (cm^3/s^2)

after which calculating ϵ is a simple matter of extracting the numerical value of the integration and scaling to convert to the cgs units used in the 1960s.

```
nu <- 1e4 * swViscosity(35,10) / swRho(35,10,10,eos=
"unesco")
15 * nu * I$value
[1] 0.6810874
```

(This ϵ estimate exceeds the Grant et al. (1962) value by 12%.)

2.6.5 Two-Dimensional Interpolation

The two-dimensional case of interpolating on a rectangular grid is handled with `interp.surface()` from the `fields` package. This does local bilinear interpolation of $z = z(x, y)$ by applying

$$(1 - x')(1 - y')z_{00} + (1 - x')y'z_{01} + x'(1 - y')z_{10} + x'y'z_{11} \tag{2.7}$$

where (x', y') is the relative position of the point within the grid cell that bounds it, and z_{00} is the value at $x' = y' = 0$, z_{01} is that at $x' = 0$, $y' = 1$, etc.

Exercise 2.40 Use `interp.surface` to find water depth H under the mean Gulf Stream position as defined in the `gs` dataset of the `ocedata` package. Draw a map of the Gulf Stream location along with a graph of how H varies with distance along the path. (See page 203 for a solution.)

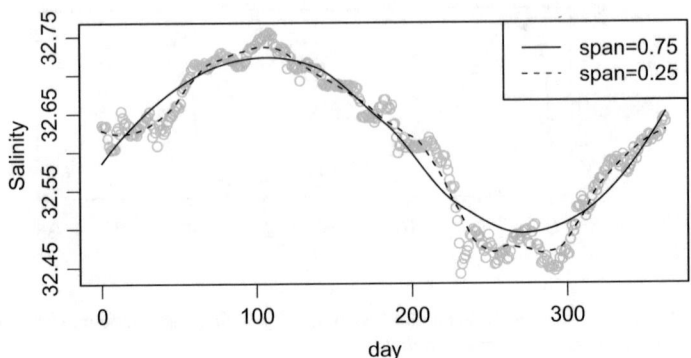

Fig. 2.26 Using `loess()` for surface salinity at ocean weather station Papa

2.6.6   Locally Weighted Polynomial Fitting

R has two functions for locally weighted polynomial fitting, `lowess()` and `loess()`. The first is used by `panel.smooth()`, which in turn is used by `plot.lm()`, `coplot()`, etc., so it deserves understanding, but the newer `loess()` is the focus here.

Figure 2.26 shows variation of surface salinity at ocean weather station Papa, plotted with

```
data(papa, package="ocedata")
day <- as.numeric(papa$t - papa$t[1]) / 86400
salinity <- papa$salinity[,1]
plot(day, salinity, ylab="Salinity", col="gray")
```

where two `loess()` fits are shown, illustrating the use of the span argument.

```
l <- loess(salinity ~ day)
lines(day, predict(l))
ll <- loess(salinity ~ day, span=0.25)
lines(day, predict(ll), lty="dashed")
legend("topright",lty=1:2,legend=c("span=0.75",
"span=0.25"))
```

2.6.7   Interpolating and Smoothing Splines

R provides interpolating splines with `spline()` and `splinefun()` and smoothing splines with `smooth.spline()`. Such functions can be quite helpful in dealing with noisy oceanographic data. For example, a smoothing spline may be fitted to the `turbulence` data with

```
s <- smooth.spline(k, k^2*phi)
```

where default values are used for additional arguments that control the degree of smoothing, knot distribution, etc.

The predicted values are found with

```
spred <- predict(s, kk)
```

producing a list with x being the k supplied to `smooth.spline()` and y being the interpolated value, so that

```
lines(spred$x, spred$y, lty="dotted")
```

could add a spline curve to Fig. 2.25.

Exercise 2.41 Contrast the predictions of interpolating and smoothing splines for the `turbulence` data. (See page 204 for a solution.)

Exercise 2.42 Create a function returning the prediction of a smoothing spline, and use it to calculate ϵ as in Sect. 2.6.4. (See page 205 for a solution.)

2.6.8 Cluster Analysis

Cluster analysis may be used to divide a set of data into subsets based on similarity of some property within groups and dissimilarity between them. Different applications call for different measures of similarity, perhaps explaining the diversity of approaches to cluster analysis (Estivill-Castro 2002).

In the popular k means clustering method (Hartigan and Wong 1979), the measure of similarity is Euclidean distance in property space, the square of which is given by

$$\sum_{i=1}^{n} (x_i - \hat{x}_i)^2 \tag{2.8}$$

where $x_1 \ldots x_n$ and $\hat{x}_1 \ldots \hat{x}_n$ are the coordinates of two points in n-dimensional property space. This has a simple interpretation if the coordinates are of an equal type, e.g. representing a geometric location, but other cases are more problematic.

For example, if x_1 is salinity and x_2 is temperature, as on a temperature-salinity diagram, then the two terms in the expanded sum within (2.8) have different units, so there can be no physical meaning to their addition. Other problems can arise even in a geometrical view, e.g. owing to the vastly different vertical and horizontal extents of oceans. For this reason, (2.8) might be better expressed in nondimensional form, as

$$\sum_{i=1}^{n} \frac{(x_i - \hat{x}_i)^2}{L_i^2} \tag{2.9}$$

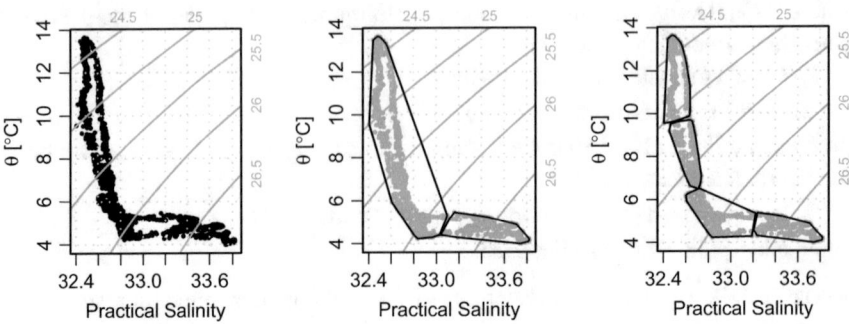

Fig. 2.27 Demonstration of cluster analysis in temperature-salinity space

where L_i is an appropriate scale in the same unit as x_i. In many practical problems, then, a core issue is the selection of the L_i values.

The methodology can be illustrated with temperature and salinity in the papa dataset, drawn in the left panel of Fig. 2.27 with.[22]

```
data(papa, package="ocedata")
S <- as.vector(papa$salinity)
T <- as.vector(papa$temperature)
p <- rep(swPressure(-papa$z), each=dim(papa$salinity)[1])
ctd <- as.ctd(S, T, p, longitude=-145, latitude=50)
plotTS(ctd, pch=20, cex=1/2, eos="unesco")
```

Most readers might describe the pattern as having two "arms", one relatively isohaline, the other relatively isothermal. However, opinions might vary, given a request to identify 10 groupings. It helps to use extra oceanographic information in deciding how many clusters to seek. Thought should also be given to the choice of independent variables. For example, in the present case, some applications might call for a transformation to density-spiciness space, which could yield different clusters. Generally, careful attention to the setup can be the key to getting sensible results from cluster analysis.

For the purpose of illustration, temperature and salinity may be used as the coordinates, with scale() used to nondimensionalize the variables, after which kmeans() may be used to do cluster analysis with 2 and 4 suggested groupings, in the middle and right panels of Fig. 2.27, with

```
plotTSCluster <- function(ctd, k=4)
{
    theta <- swTheta(ctd)
    Stheta <- scale(cbind(S, theta), TRUE, TRUE)
    cl <- kmeans(Stheta, k, nstart=30)
```

---

[22]It is not strictly necessary to use as.ctd() to create a "ctd" object, but it makes it easier to create a standardized plot with isopycnals.

```
plotTS(ctd, col="darkgray", pch=20,cex=0.5,
eos="unesco")
which <- cl$cluster
for (i in 1:k) {
    x <- S[which==i]
    y <- theta[which==i]
    hull <- chull(x, y) # chull() computes
    complex hulls
    hull <- c(hull, hull[1])
    lines(x[hull], y[hull])
  }
}
set.seed(268)                    # for reproducibility
plotTSCluster(ctd, 2)
plotTSCluster(ctd, 4)
```

in which chull() has been used to find the convex hull polygons surrounding the clusters. Note also that kmeans() uses random numbers in its search for cluster centres, so repeating a calculation without using set.seed() can yield different answers in some cases.

2.6.9 Fast Fourier Transforms

The fft() function, which provides forward and inverse fast Fourier transforms (FFT), is used by convolve() and spectrum(), and it is also useful by itself. It does not normalize in either direction, leaving this to the user, e.g.

```
fftn <- function(z, inverse=FALSE)
        fft(z, inverse) / sqrt(length(z))
```

defines a 1-D wrapper with a common normalization (see Exercise 5.25 for an application to rotary spectra.)

With fftn() thus defined, a test of Parseval's Theorem might be, e.g.

```
library(testthat)
x <- rnorm(100)
X <- fftn(x)
xx <- fftn(X, TRUE)
expect_equal(sum(x^2), sum(Mod(X)^2))
```

where the handy expect_equal() from the testthat package is used to check that the time-series variance matches the integral of the power spectrum. (If the test failed, it would print an error message.)

A check of the invertibility of the fftn() formulation might be

```
expect_equal(x+0i, fftn(fftn(x), inverse=TRUE))
```

2.7   Input and Output

R has sufficient flexibility to handle any conceivable file format. Since this is achieved by a fairly long list of functions and arguments, the present discussion covers several pages, despite being a thin summary.

2.7.1   Reading from Text Files

2.7.1.1   Simple Tables

If a file named `table_eg_1.dat` starts with

```
x y
1 1.6180
2 2.7183
3 3.1416
```

then

```
d <- read.table("table_eg_1.dat", header=TRUE)
```
will read the data into a data frame (see Sect. 2.3.8) named d. Since `header` is `TRUE`, the column names are inferred from the first line of the file. If the `header` argument is dropped, the columns will be named `V1` and `V2`, unless the `col.names` argument is supplied. This convention applies also to relatives such as `read.csv()`, `read.fwf()` and `read.fortran()` in the `utils` package, and analogues in the `readr` package, which can be faster.[23]

2.7.1.2   Complicated Tables

A Southern Oscillation Index (SOI) dataset[24] starts as follows:

```
1866 -1.2 -0.3 -1.0 -0.7   0.1 -0.9 -0.7   0.7 -0.4   0.1
        1.6 -0.3
1867   0.4 -0.0 -0.0   0.8   0.7 -0.5   0.6   0.5   0.1 -0.7
       -1.2 -1.7
```

One way to read such data would be to create a matrix

```
d <- as.matrix(read.table("../data/soi.dat", header=FALSE))
```
and then construct a decimal-year time vector

---

[23] A test with a 90 Mb file on the author's machine revealed `read_csv()` to be nearly 6 times faster than `read.csv()`.

[24] http://www.cgd.ucar.edu/cas/catalog/climind/SOI.signal.ascii.

```
y <- d[,1]
year <- seq(from=head(y,1), to=tail(y,1)+11/12, by=1/12)
```
The first step in isolating the SOI values is to drop the first column of d, after which the matrix should be transposed with t() before creating the vector:
```
soi <- as.vector(t(d[,-1]))
```
It is necessary to account for missing values, equal to -99.9 in this dataset, but it is risky to check for numerical equality, so
```
missing <- soi < (-90) # parentheses prevents "<-" typo
```
may be preferable. Usually, the next step would be to set the missing values to the R coded value, with soi[missing] <- NA, but with this particular dataset, the missing values are all at the end,
```
year <- year[!missing]
soi <- soi[!missing]
```
might be used, if there were no need to retain data length.

2.7.1.3 Line-Based Input

The readLines() function reads a file by lines, returning a vector of strings, one per line. To illustrate how this can be useful in decoding complicated data files, a mapping application was used to generate an automobile route from Dalhousie University to the Woods Hole Oceanographic Institution, resulting in an XML file. In this file, geographical locations are delimited by strings <coordinates> and </coordinates>. Within these blocks are comma-separated values of longitude, latitude and a third item.

The first step in inferring the route is to read the whole file as strings, and to find the portion containing the route
```
d <- readLines("../data/dalwhoi.kml")
start <- grep("^\\ s*<coordinates>\\ s*$", d)
end <- grep("^\\ s*</coordinates>\\ s*$", d)
```
Here, ^ stands for the string start, \\s* stands for whitespace, and $ stands for the string end (see Sect. 2.3.3.3). Next,
```
pathIndices <- seq(start + 1, end - 1)
data <- read.csv(text=d[pathIndices], header=FALSE)
lon <- data$V1
lat <- data$V2
```
reads the locations. Figure 2.28 is then constructed with
```
data(coastlineWorldFine, package="ocedata")
mapPlot(coastlineWorldFine, projection="+proj=merc",
        col="lightgray", longitudelim=range(lon),
        latitudelim=range(lat))
mapLines(lon, lat, lwd=5) # thick line for the route
```
which makes use of a Mercator projection (see Chap. 3).

Fig. 2.28 Road from
Dalhousie University to
Woods Hole Oceanographic
Institution

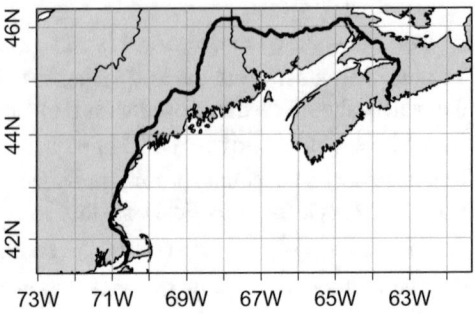

Exercise 2.43 Read the Dalhousie-WHOI route using the XML package. (See page 205 for a solution.)

Exercise 2.44 Read the sample CTD file ctd.cnv, skipping the header and naming the columns. (See page 206 for a solution.)

2.7.1.4   Reading by Words and Characters

The scan() function reads a file a "word" (defined as a character group-ing separated by whitespace) at a time, returning the vector of items, e.g. if ../data/soi.dat contains the SOI data,[25] then

```
tokens <- scan("../data/soi.dat")
length(tokens)
[1] 1963
```

reveals that scan() reads the whole file, finding 1963 words.

At a finer level, readChar() can be used to read individual characters, e.g.

```
readChar("../data/soi.dat", 10)
[1] " 1866   -1."
```

Exercise 2.45 Read the SOI data with scan(). (See page 207 for a solution.)

2.7.1.5   File and Text Connections

When supplied with a filename, most reading functions that are directed at files start afresh with each invocation, e.g.

```
readChar("../data/soi.dat", 30)
[1] " 1866   -1.2   -0.3   -1.0   -0.7 "
readChar("../data/soi.dat", 30)
[1] " 1866   -1.2   -0.3   -1.0   -0.7 "
```

---

[25]http://www.cgd.ucar.edu/cas/catalog/climind/SOI.signal.ascii.

shows that `readChar()` rereads the first 30 characters each time it is called. File connections provide a way to read sequentially through files. The value returned by `file()` is a file connection, and e.g.

```
soi <- file("../data/soi.dat", "r") # second arg
means read-only
readChar(soi, 15)
| [1] " 1866  -1.2  -0"
readChar(soi, 15)
| [1] ".3  -1.0  -0.7 "
```

demonstrates that `readChar()` retains a pointer to file location when its first argument is a file connection, as opposed to a file name.

Connections may also be made to strings, using `textConnection()`, e.g.

```
con <- textConnection("but not a drop to drink")
scan(con, "character", nmax=3)
| [1] "but" "not" "a"
scan(con, "character", nmax=3)
| [1] "drop"  "to"    "drink"
```

Readers with programming experience are likely to see that connections are the key to working with complex files such as are common in oceanography.

2.7.2 Reading Binary Data

R has powerful and flexible tools for working with binary files. This is important for oceanographic work, because many instruments record in binary format. Working with binary data is a somewhat complicated business in any language, but readers with some skill (e.g. those who know the meaning of phrases such as "little endian" and "unsigned int") should not have difficulty handling their data in R.

The first processing step is typically to read the entire file into a memory buffer, after which the buffer is examined in detail. For example, the `oce` package (Chap. 3) reads acoustic Doppler files with code of the form

```
file <- file(filename, "rb")
seek(file, 0, "end")
fileSize <- seek(file, 0, "start")
```

Here, `file()` is given argument `"rb"` to open the file read-only, in binary format. Then `seek()` is used to "point" to the end of the file. Calling `seek()` a second time moves the pointer to the start of the file and also returns the number of bytes in the file, saved here as `fileSize`. Now, the whole file may be read into a buffer with `readBin()`, in binary ("raw") form.

```
buf <- readBin(file, "raw", fileSize)
```

Each element of `buf` corresponds to a byte. A common operation is to check those bytes for coded sequences that flag data sections. For example, ADCP files from RDI-Teledynestart with the byte `0x7f` repeated twice, so a test to see if the file might be of this type is

```
probablyRdiAdcp <- buf[1] == 0x7f && buf[2] == 0x7f
```

and this is one of many tests used by `oceMagic()` to infer file type, helping `read.oce()` to select a specialized reading routine (Sect. 3.2).

Another common operation is to match bytes for sequences that occur at arbitrary locations within files, not just at the start. In many files, data are provided in discrete chunks, with each one starting with a particular byte sequence. A single byte is not a good flag for this, because the probability of a random byte matching is 1/256, which is so high that typical files will have many false positives. This explains why the scheme is usually to have a multi-byte sequence, along with other clues, such as checksums, that identify data chunks.

For large or complex files, it can prove convenient to switch from R to C or C++, to gain processing speed and to simplify programming. For example, the calculation of checksums in acoustic Doppler files is relatively simple in C or C++, each of which is very well-suited to computation at the byte level (Fig. 1.1). The efficiencies gained by moving some core calculations to such compiled languages are dramatic. Since these are not languages known to all oceanographers, it is convenient that `oce` handles many important file types, freeing analysts from the need to go beyond R in day-to-day work.

2.7.3   Reading Databases

R can read and write several types of database, e.g. mySQL and SQLite are handled with the `RMySQL` and `RSQLite` packages. The interfaces are similar enough that a single illustration should suffice.

In 2011, the author started a community educational project to monitor sky light using data loggers that measure light levels and feed the results to a collating computer. This collating computer uses an SQLite database named `skyview.db` that was originally created with

```
CREATE TABLE observations(id integer primary key,
                          time int, light_mean real);
```

where `time` is the observation time in Unix seconds and `light_mean` is the mean light level during a sampling interval.

The collating computer updates graphs on webpages on a regular interval, using code that starts with

```
library(RSQLite)
m <- dbDriver("SQLite")
d <- dbConnect(m, dbname="../data/skyview.db")
```

to load the SQLite driver and connect to the database. Then, it is a simple matter of querying the database to extract the columns from the table. For example, the `observations` table is recovered with

```
o <- dbReadTable(d, "observations")
```

which creates o as a data frame including all the data in the observations table. If the goal were to extract only certain columns, a query such as

```
o<-dbGetQuery(d, "select time, light_mean from
observations")
```

might be used. (The second argument to dbGetQuery() is written in SQL, which can be an advantage in working with a large database that can be pared down at a low level, before transmission of data to R.)

Exercise 2.46 From https://rbr-global.com/support/matlab-tools, get RSKtools and extract time and temperature from the SQLite file named sample.rsk. (See page 207 for a solution.)

2.7.4 Reading NetCDF Files

The NetCDF file format is popular in oceanography and meteorology, particularly for data that can be expressed as vectors or arrays. Of its several useful features, two that stand out are its use of an endian-independent binary format and its tight binding of data with metadata that indicate units, ranges, data-quality flags, etc.

An example is provided by the 5-degree resolution version of the World Ocean Atlas (see, e.g., Boyer et al. 2009), available from NODC.[26] This provides gridded values of salinity and temperature, etc. Accessing the data is easy with the ncdf4 package, written by David W. Pierce. The first step is to open a file connection, e.g. with

```
library(ncdf4)
con <- nc_open("../data/woa13_decav_t00_5dv2.nc")
```

after which an overview of the contents is found by printing con. The results, which are detailed and much too long to show here, indicate that the file contains coordinate variables named lon, lat and depth, in addition to arrays holding temperature information in different forms. Data may be accessed with ncvar_get(), e.g. the atlas grid geometry is recovered with

```
lon <- ncvar_get(con, "lon")
lat <- ncvar_get(con, "lat")
depth <- ncvar_get(con, "depth")
```

and a 3D array containing analysed temperature is recovered with

```
t_mn <- ncvar_get(con, "t_mn")
```

With such NetCDF-specific work complete, it is easy to explore the data. For example, Fig. 2.29 shows latitudinal dependence of sea-surface temperature, created with

---

[26]https://www.nodc.noaa.gov/OC5/woa13/woa13data.html.

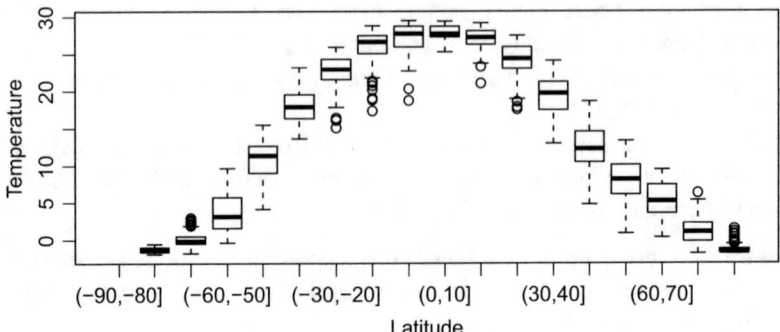

Fig. 2.29 Variation of sea-surface temperature with latitude and longitude

```
SST <- t_mn[, , depth==0]           # third index is depth
vSST <- as.vector(SST)
vlat <- rep(lat, each=length(lon))
lat10 <- cut(vlat, breaks=seq(-90, 90, 10))
boxplot(vSST ~ lat10, xlab="Latitude",
ylab="Temperature")
```

where cut() has been used to break latitude up into 10-degree bands. Note the casting of the temperature matrix into vector form with as.vector(), and the creation of a corresponding latitude by replication with rep().

Exercise 2.47 Plot SST contours with a coastline. (See page 207 for a solution.)

2.7.5 Writing Files

R provides a variety of methods for writing to files. This may be done in textual and binary ways. Conveniently, the names of functions for writing tend to be paired with those for reading, e.g. writeChar() is a sibling to readChar().

The R.matlab package can be used to write in Matlab format, but the reader is cautioned (as of the writing of this book) that the package is somewhat limited, e.g. for arrays with more than 2 dimensions.

An important R function for writing is save(), which stores R objects to a file in a binary format that preserves their structure and contents, suitable for recovery with load(). Saving data in this way offers a convenient way to buffer results, avoiding repeating slow calculations across R sessions. Importantly, save() uses a binary format that retains full numerical resolution, alleviating any need to decide how many digits to use in a text-based file. Since save() and load() store endian information, there is no need to worry about spurious results in transferring data between machines of different architectures.

2.8   Creating R GUI Systems

As mentioned in Sect. 2.4.15, R can handle graphical input in several ways. A simple method is to use `locator()` to find the location of a user's mouse click on a plot, e.g. to take advantage of the fact that an analyst might be able to detect a data anomaly more easily by examining a graph than by devising a new algorithm.

Such approaches are helpful for simple tasks, but more sophisticated GUI elements are desirable for complex tasks, and that's where the `shiny` package comes in. This package makes it easy to set up GUI systems (or "apps") in R. It has tools for creating GUI elements such as sliders, menus, radio boxes, text boxes, etc., in addition to a powerful and flexible system for connecting such elements to general R code. For example, a `shiny` app can be set up so that radio buttons control whether to represent data with contours or images, with slider bars to control parameters used in smoothing data, etc. The `shiny` system also has good support for mouse clicks and drags within a plotting area, which can be helpful in selecting data to flag or subregions to be replotted at higher magnification. Importantly, the user's actions can be accessed throughout the app, which means that a user action in one panel of a plot can control the display in another panel, providing an escape from the focus limitation of multi-panel plots that was mentioned in Sect. 2.4.15.

There are many online tutorials dealing with `shiny`, those provided by `Rstudio` being noteworthy.[27] A good way to learn `shiny` is to download a tutorial app and start modifying it. For example, a common way to search for errors in CTD data is to look for outliers on a $T–S$ plot. A plot-interaction-exclude app provided on the `Rstudio` website[28] addresses a similar task, and so it provides a starting point for a CTD-editing app.

It can be helpful to divide app code into two files, one defining the user interface and the other defining the actions associated with the interface elements. For the former, saving

```
library(shiny)
verticalLayout(wellPanel(plotOutput("plot", brush="brush")),
               wellPanel(actionButton("toggle", "Toggle"),
                         actionButton("reset", "Reset"),
                         textOutput("excluded")))
```

to a file named `ui.R` sets up an interface with a plot area ("`wellPanel`" in `shiny` parlance) at the top of the window and an action area below, as illustrated in Fig. 2.30. The user is invited to drag the mouse to select ("brush") bad data, and then click "Toggle" to switch the status of the indicated points from good to bad (and vice-versa, for corrections). The other items of the second `wellPanel()` will be used to list the indices of suspicious data, and to provide a way to reset the analysis.

---

[27] https://www.rstudio.com/products/shiny.

[28] http://shiny.rstudio.com/gallery/plot-interaction-exclude.html.

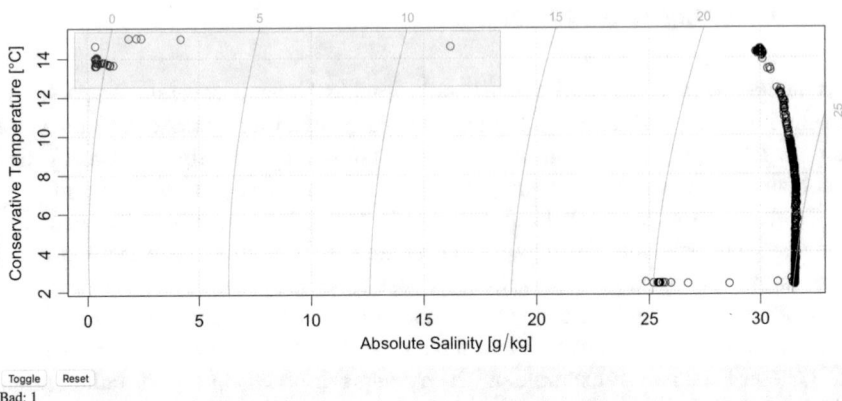

Fig. 2.30 Using the CTD-edit shiny app. A point with spurious salinity of 99 has already been designated as bad, and eliminated from the plot. The mouse has been dragged over low-salinity values that are suspect, and these will disappear when "Toggle" is pressed

These actions are accomplished and demonstrated using a built-in dataset with the following code, saved in a file named server.R.

```
library(shiny)
library(oce)
data(ctdRaw)
ctd <- ctdRaw
shinyServer(function(input, output) {
  n <- length(ctd[["pressure"]])
  vals <- reactiveValues(keep=rep(TRUE, n))
  output$plot <- renderPlot({
    plotTS(subset(ctd, vals$keep), eos="gsw")
    discard <- subset(ctd, !vals$keep)
    points(discard[["SA"]], discard[["CT"]], pch=20,
    col="red")
  })
  output$excluded <- renderText({
    paste("Bad:", paste(which(!vals$keep),
    collapse=" "))
  })
  observeEvent(input$toggle, {
    df <- data.frame(SA=ctd[["SA"]], CT=ctd[["CT"]])
    res <- brushedPoints(df, input$brush,
                         "SA", "CT", allRows=TRUE)
    vals$keep <- xor(vals$keep, res$selected_)
  })
```

```
observeEvent(input$reset, {
    vals$keep <- rep(TRUE, n)
  })
})
```

In examining such code, the reader would benefit from reading the documentation for `reactiveValues()` and `observeEvent()`, for these are key tools used to control interactions throughout the app.

Exercise 2.48 Extend the CTD-edit `shiny` app to read data from a file, set flags for bad data and save the result to a file. (See page 208 for a solution.)

2.9 Debugging

Debugging tools are an important part of any computing system, and a good topic with which to end this tutorial. R comes with a simple but effective set of debugging tools that should become familiar to any user who writes programs of significant complexity.

Debugging is best done in an interactive environment. The present discussion describes actions that can be taken in the R console or in an editor, but readers who are using the `Rstudio` system will find that there are GUI actions for the procedures outlined here.[29] If `file.R` produces an error when run from the (presumed Unix) operating system with

```
Rscript file.R
```

then the first step is to launch an interactive R environment, and execute

```
source("file.R")
```

to replicate the problem. Then, using

```
traceback()
```

will reveal the spot where the error occurred, along with an execution trace indicating how control arrived at that spot. If the error is obvious, `file.R` can be altered and re-sourced, to work through whatever solution occurs to the coder. It is also helpful to set the system up to trigger the debugger upon errors.

Sometimes an error will be reported long after something was done incorrectly. (For example, line 100 might contain an attempt to read a file that was incorrectly named in line 50.) For this reason, it can be helpful to add lines that print variables at strategic spots in the code. This can be time consuming, so a better approach can be to use `browser()` or `debug()`. For example, with error at line 200, a good approach might be to add

```
browser()
```

---

[29]RStudio has a variety of other helpful features, e.g. a code-completing editor and a code-analysis tool that can recommend alterations that may make code more robust.

at line 199. When R gets to that spot, the prompt will change, and the user will be able to examine any variable of interest. Actually, any R command can be executed, so the user is free to try to find the problem by plotting, etc. There are also some control commands that can be executed in a browser() session, the most useful of which are n (or an empty line), which advances to the next step of execution, and Q, which exits the browser. Although it is simple, browser() is tremendously helpful in finding errors of any kind, and it may be the most important debugging tool a R programmer can learn.

Sometimes it is not desired to edit the code, so inserting calls to browser() is not an option. For this purpose, debug() is handy. It takes as its first argument the name of a function. Then, when execution is started, everything proceeds normally until that function is encountered, whereupon control is handed over to browser(). It is also possible to instruct Rstudio to act as though a break() call had been been inserted at a specified spot in the code, without actually modifying the source file.

In the special case of the oce package, another debugging method is to set the debug argument to an integer higher than zero, so that functions print a record of some important aspects of their processing. This is just one of the many practical aspects of the oce package that are sketched in the next chapter.

Chapter 3
The oce Package

Abstract The oce package simplifies oceanographic analysis by handling the details of discipline-specific file formats, calculations and plots. Designed for real-world application and developed with open-source protocols, oce supports a broad range of practical work. Generic functions take care of general operations such as subsetting and plotting data, while specialized functions address more specific tasks such as hydrographic analysis, ADCP coordinate transformations, etc. It is easy to document work done with oce, because its functions automatically update processing logs stored within its data objects. Users are not limited to oce functions, however; data are extracted easily from oce objects, so that the thousands of other R packages may be used as needed.

3.1 Package Options

The oce package (Kelley and Richards 2018) has several options to control global behaviour, e.g.

```
options(oceDebug=3)
```

sets debugging to a high level for all oce function calls, as an alternative to setting debug argument in such calls. Some other options are listed in Table 3.1. It is typical to set such things in a startup file (see Sect. 2.2.4).

3.2 File Formats

Table 3.2 lists some of the data formats recognized by read.oce(), which uses oceMagic() to infer file type, and then calls a specialized function to read the data. These specialized functions may also be called directly. Either way, users are relieved from reading lengthy data-format specifications and writing complex

Table 3.1 Some user-controllable oce startup options

Option	Default value	Meaning
oceMar	c(3, 3, 2, 2)	Value for par(mar), controlling margin widths
oceMgp	c(2.0, 0.7, 0)	Value for par(mgp), controlling axis label locations
oceDrawTimeRange	TRUE	Should oce.plot.ts() show the time range?
oceAbbreviateTimeRange	TRUE	Should oce.plot.ts() shorten time ranges?
oceTimeFormat	"%Y-%m-%d %H:%M:%S"	Format for time strings
oceUnitBracket	"["	Character to embrace units in plot labels; can also be "("
oceEOS	"unesco"	Preferred seawater equation of state; can also be "gsw"; see Sect. 5.2.1 and Appendix D

Table 3.2 Some of the oceanographic data formats recognized by read.oce() and its helper function, oceMagic

Class	Details
adp	Acoustic Doppler profiler, in RDI-Teledyne, Nortek or Sontek format
adv	Acoustic Doppler velocimeter, in Nortek or Sontek format
amsr	AMSR satellite data
argo	Argo float data
bremen	Data format used at Bremen
cm	Current meter, in Interocean format
coastline	Coastline shape, in mapgen, shapefile and other formats
ctd	CTD, in Seabird *.cnv, WOCE exchange, ODF or Ruskin format
echosounder	Biosonics scientific echosounder
g1sst	Global 1km SST satellite/model data
gps	Location data
ladp	Lowered Acoustic Doppler profiler
landsat	Landsat satellite data
lisst	Laser in situ scattering and transmissometry
lobo	Land/Ocean biogeochemistry Observatory
met	Meteorological data.
oce	Base of all classes in the oce package
odf	Data format used by Department of Fisheries and Oceans, Canada
rsk	RBR logging devices, e.g. temperature-depth recorders
satellite	Base of amsr, g1sst and landsat classes
sealevel	Sea-level elevation, in MEDS or Hawaii format
section	Section data
tidem	Tidal-model data
topo	Earth topography, in NOAA format
windrose	Wind rose data

code,[1] e.g. the specialized graphical representation of CTD data shown in Fig. 3.1 was constructed with a simple call to `plot()`, which tailors its action to the class of its first argument.

The ability to read a wide variety of data types is a good reason to try R and oce for oceanographic analysis. There are two main advantages over software provided by manufacturers. First, oce is open-source, and thus easy to inspect or modify. Second, manufacturers provide software for just their own instruments, which is of limited help in coordinating data from the typical oceanographic experiment, which employs a variety of instrument types.

Open-source alternatives are available in Matlab and Python, and readers will likely find themselves using these from time to time. A weakness of many such systems is that they tend to be specialized to particular instruments. By contrast, oce handles many instruments in a uniform way, which can be helpful to analysts who work with several data types at once. Much of the oce uniformity stems from its object-orientation design, discussed in the next section.

3.3 Object Orientation

The oce package uses the S4 scheme of object orientation[2] with a hierarchical collection of object classes that inherit from a common base class named "oce". This inheritance scheme simplifies the internal coding of oce, reducing the chance of bugs and also making it easier for users to add new objects.

All oce classes have three S4 "slots," with contents as follows (Fig. 3.2).

* metadata, a list describing the object. The contents vary with the object type, perhaps including the name of a data file, the sampling location, etc.
* data, a list containing the actual data. Again, the contents depend on the object. For example, CTD objects contain vectors for hydrographic quantities, ADP objects contain vectors for time and distance in addition to arrays for velocity components, etc. (This combination of vectors and arrays explains why a list is used instead of a data frame; see Sect. 2.3.6.)
* processinglog, a list containing items named time and value that record the processing steps that led to present state of the object.

There are two ways to access data within oce objects. It is possible to use the @ symbol to access information stored in an object's data or metadata slots. However, the recommended method is to the access operator " [[", e.g.

---

[1] The effort of decoding oceanographic data files can be significant, e.g. Teledyne-RDI (2007) devotes nearly 30 pages to byte-level format of ADCP files, and following that format requires several hundred lines of R and C/C++ code.

[2] There is no need to understand S4 in order to use oce, but curious readers can get the gist from `help(Classes)` or Chapter 9 of Chambers (2008).

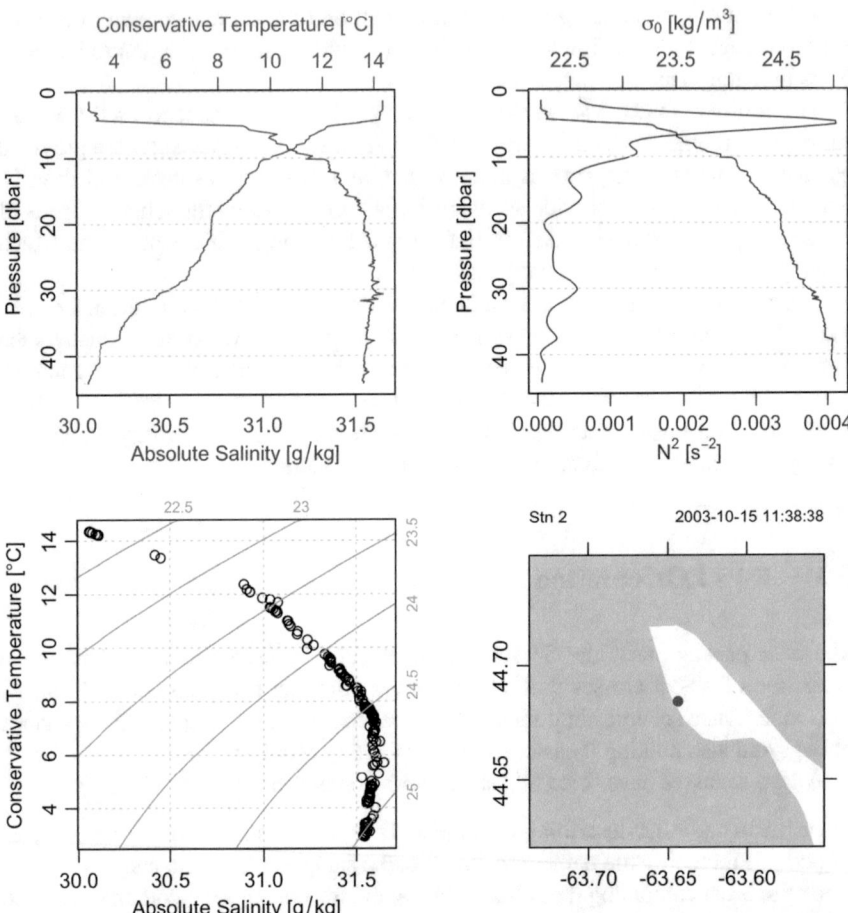

Fig. 3.1 Hydrographic diagram of a CTD cast made in Halifax Harbour by students in the author's Physical Oceanography class at Dalhousie University. This diagram was produced with just two oce function calls: one to read the data, another to plot them

```
data(ctd, package="oce")
head(ctd[["temperature"]])
[1] 14.22109 14.22649 14.22509 14.22219 14.22669
14.23318
```

There are two advantages of the accessor approach. First, it isolates users from the details of internal storage, letting users write code that is resistent to any changes in the internal structure of oce objects that may be necessitated by changes in instrumentation or analysis methodologies. Second, accessors make it easy for users to infer derived quantities that are not actually stored in the data object, e.g. potential temperature for a CTD or attenuation-corrected backscatter strength for an ADCP.

The " [[" operator works for assignment as well as access, e.g. temperature might be increased with

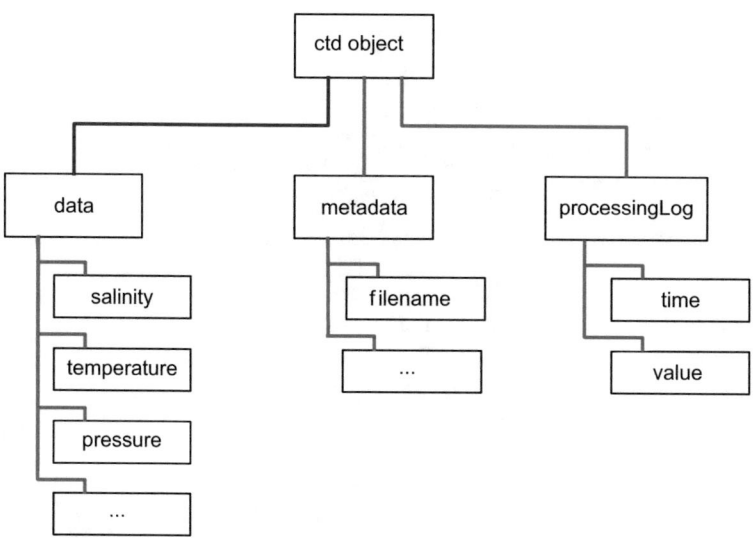

Fig. 3.2 Structure of a CTD object

```
ctd[["temperature"]] <- 0.001 + ctd[["temperature"]]
```
but this scheme should be used only for quantities actually stored in the object, not for derived quantities. (For nontrivial changes, however, it is recommended to use `oceSetData()` and `oceSetMetadata()`, since these record changes within the object's processing log.)

Internally, "`[[`" is a function that searches through the object for the named quantity. It first examines the `metadata` slot, returning the value (for access or assignment) if found, otherwise moving on to the `data` slot and repeating the test. This scheme permits a uniform notation, no matter which slot holds the information. This is useful because the appropriate slot depends on the object class, e.g. `latitude` is mandatory for a `coastline` object so it belongs in the `data` slot, but it is an optional addition for CTD instruments, so it belongs in the `metadata` slot. The code fragment `a[["latitude"]]` works for both types of object, taking values from different places depending on the class of a. This scheme makes it easy for an analyst to work with a wide range of data types without case-by-case tailoring of code.

3.4 Datasets

Several datasets are provided with `oce` and `ocedata`, some of which are listed in Tables 3.3 and 3.4. The dataset documentation can be a useful adjunct to the help on its related class; e.g. compare the output of `help("ctd")` with the more detailed information that `help("ctd-class")` provides.

Table 3.3 Datasets provided in the oce package

Name	Description
adp	SLEIWEX ADCP measurements
adv	SLEIWEX ADV measurements
argo	Argo float #3900388 measurements
cm	SLEIWEX S4 current meter measurements
coastlineWorld	Default (1:50M) world coastline
colours	Colours used in some oce palettes
ctd	CTD profile collected in Halifax Harbour
ctdRaw	Raw CTD data, including calibration and upcast
echosounder	SLEIWEX echosounder measurements
landsat	Data from a Landsat image
lisst	LISST dataset, constructed artificially
lobo	LOBO measurements made in Halifax Harbour
met	Meteorological observations at Halifax Int'l Airport
rsk	SLEIWEX temperature-depth recorder data (RBR logger)
sealevel	Sea-level variation within Halifax Harbour during 2003
sealevelTuktoyaktuk	Sea-level variation near Tuktoyaktuk, from Foreman (1977)
section	WOCE hydrographic section designated A03
tidedata	Data on tidal constituents, used by tidem()
topoWorld	World topography data on a 12-minute grid
wind	Wind data in Koch et al. (1983)

3.5 Functions

The oce package provides generic functions (Sect. 2.3.11.6) to handle common tasks, including the access operator mentioned above, along with subset(), summary() and plot(). This scheme lets users ignore the internal structure of the data, e.g. if d is an oce object including time, then

```
dd <- subset(d, time < mean(range(d[["time"]],
na.rm=TRUE)))
```

retrieves data from the early portion of the sampling interval, no matter the object's class. Plotting is also done with generic functions, e.g. Fig. 3.1 was produced with

```
data(ctd, package="oce")
plot(ctd)
```

where the plot details are obtained with either of the following:

```
help("plot,ctd-method")
?"plot,ctd-method"
```

In addition to generic functions, oce provides a long list of functions for specialized oceanographic tasks, including (with * representing several function names)

- map*() functions for drawing maps with projections (see Sect. 3.6)

Table 3.4 Datasets provided in the `ocedata` package

Name	Description
RRprofile	Hydrographic profile from Reiniger and Ross (1968)
beaufort	A CTD profile in the Beaufort Sea
buoy	Measurements made by a buoy off Halifax
coastlineWorldFine	Fine-resolution (1:10M) world coastline
coastlineWorldMedium	Medium-resolution (1:50M) world coastline
conveyor	Some points on the Broecker (1991) "conveyor belt"
drag	Air-sea drag coefficients from Garratt (1977)
endeavour	Path of HMS Endeavour
geosecs235	GEOSECS tritium station 235
giss	Goddard Institute for Space Studies temperature timeseries
gs	Gulf Stream position, from Drinkwater et al. (1994)
levitus	"Levitus" World Ocean Atlas SSS and SST
munk	Pacific temperature profile examined by Munk (1966)
nao	North Atlantic Oscillation timeseries
oceans	Geometry of some oceans
papa	Measurements at Ocean Weather Station P
redfieldNC	Nitrate-carbon data in Figure 3 of Redfield (1934)
redfieldNP	Nitrate-phosphate data in Figure 1 of Redfield (1934)
redfieldPlankton	Plankton data in Table II of Redfield (1934)
riley	Plankton data in Figure 21 of Riley (1946)
schmitt	Temperature-salinity data in Figure 1 of Schmitt (1981)
secchi	Sechhi-disk measurements in North and Baltic Seas
soi	Southern Oscillation Index from 1866
topo2	World topography data on a 2-degree grid
turbulence	Turbulence measurements by Grant et al. (1962)
wilson	Seafloor-spreading data in Table 1 of Wilson (1963)

- `oce.plot.ts()`, an alternative to `plot.ts()` for time-series data
- `imagep()` and `drawPalette()` for colour palettes in images and generally
- `pwelch()` for averaged spectra as discussed by Welch (1967)
- `atm*()` functions relating to atmospheric properties
- `sw*()` functions relating to seawater properties (see Table 3.5 and Sect. 5.2.1)

Exercise 3.1 Use the generic `plot()` for CTD objects, to produce a version of Fig. 3.1 using the UNESCO equation of state instead of the default TEOS-10 version. (See page 209 for a solution.)

Exercise 3.2 (a) Calculate the density of seawater at pressure 100 dbar, salinity 34 PSU, and temperature 10 °C. (b) What temperature would the parcel have if raised adiabatically to the surface? (c) What density would it have if raised adiabatically to the surface? (d) What density would it have if lowered about 100 m,

Table 3.5 Some functions relating to seawater properties

Function	Description
swAbsoluteSalinity()	Absolute salinity, S_A
swAlpha()	Thermal expansion coefficient, $\alpha = -\rho_0^{-1}\partial\rho/\partial T$
swAlphaOverBeta()	Ratio of thermal and haline coefficients, α/β
swBeta()	Haline contraction coefficient, $\beta = \rho_0^{-1}\partial\rho/\partial S$
swConductivity()	Electrical conductivity, C
swConservativeTemperature()	Conservative temperature, Θ
swDepth()	Depth, $-z$, inferred from p and latitude
swDynamicHeight()	Dynamic height
swLapseRate()	Adiabatic lapse rate
swN2()	Square of buoyancy frequency, N^2
swRho()	Density, $\rho = \rho(S, T, p)$
swSCTp()	S inferred from conductivity, T and p
swSTrho()	S inferred from T and ρ
swSigma()	$\sigma = \rho - 1000\,\text{kg/m}^3$
swSigmaT()	$\sigma(S, T, 0)$
swSigmaTheta()	$\sigma(S, \theta, 0)$
swSoundAbsorption()	Sound absorption
swSoundSpeed()	Sound speed
swSpecificHeat()	Specific heat
swSpice()	Spiciness, a property orthogonal to density
swTFreeze()	Freezing temperature
swTSrho()	T inferred from S and ρ
swTheta()	Potential temperature, θ
swViscosity()	Dynamic viscosity, μ
swZ()	Vertical coordinate, z, inferred from p and latitude

Here, C represents electrical conductivity, p pressure, S salinity and T in situ temperature

increasing the pressure to 200 dbar? (e) Draw a blank T-S diagram with S from 30 to 40 PSU and T from -2 to 20 °C. (See page 209 for a solution.)

Exercise 3.3 Use propagate from the propagate package to estimate typical CTD salinity uncertainty. (See page 209 for a solution.)

3.6 A Practical Example

Figure 3.3 shows annual-mean world sea-surface temperature (SST) from the 2009 version of the World Ocean Atlas dataset (Locarnini et al. 2010; Antonov et al. 2010). A detailed explanation of the construction this diagram provides the chance to highlight some important oce functions. The first step is to access the data,

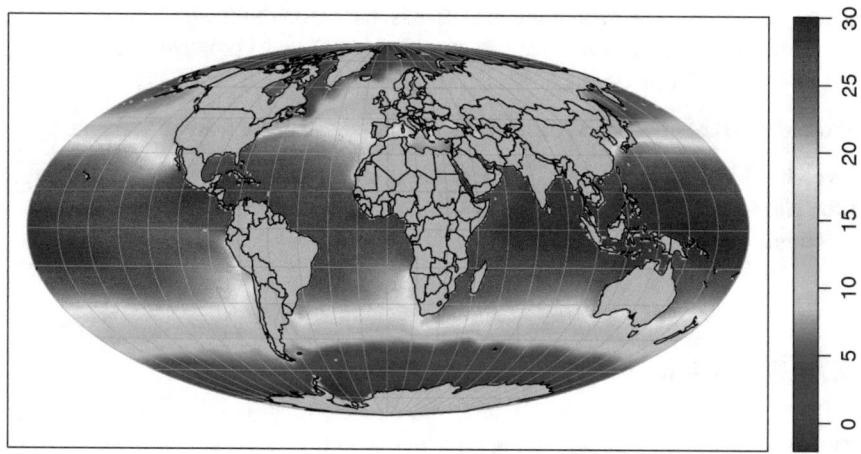

Fig. 3.3 Annual-mean sea surface temperature shown in Mollweide projection

a convenient (but spatially coarse) form of which is provided by the `ocedata` package:

```
data(levitus, package="ocedata")
```

Although `oce` can easily select a colour-scale for the image, analysts usually prefer to set such things to achieve uniformity across plots, and this may be done with

```
cm <- colormap(zlim=c(-2, 30), col=oceColorsJet)
```

which uses the "jet" color mapping (see Sect. 2.4.14). Then a palette is drawn with

```
drawPalette(colormap=cm)
```

At a global scale, the coastline provided with `oce` provides sufficient detail and the Mollweide projection may be a good choice (see Appendix C for more on projections); with these choices,

```
data(coastlineWorld, package="oce")
mapPlot(coastlineWorld, projection="+proj=moll",
grid=FALSE)
```

draws the gray land area in Fig. 3.3. Finally,

```
mapImage(levitus$longitude, levitus$latitude,
          levitus$SST, colormap=cm)
```

adds the sea-surface temperature. Readers who are following along will notice that some of the image grid elements are painting over the land. This problem is alleviated by redrawing that land, after first drawing lines of longitude and latitude:

```
mapGrid()
mapLines(coastlineWorld)
```

thus completing Fig. 3.3.

Readers might wish to examine the documentation of the relevant functions to understand this example fully, but the above should indicate the potential of `oce` to produce useful specialized oceanographic plots, in addition to those offered by its generic functions. The main thing to realize is that oce is built with R base graphics, which means that a painting model is employed, with new graphical elements being put on top of existing ones.

Exercise 3.4 Map ocean-surface density. (See page 210 for a solution.)

Exercise 3.5 Use mapPlot() to draw a world coastline with the Robinson projection, and trace the 1700s H.M.S. Endeavour cruise. (See page 210 for a solution.)

3.7 Evolution of oce

The oce package began with ad hoc code to read CTD files stored in the ".cnv" format (see Exercise 2.44). This was a main program that consisted of little more than a call to read.table() to read a specified file, with the value of the skip argument chosen after inspection of the lines at the start of that particular file.

Headers in .cnv files are of variable length, and it is tedious to alter skip for each case, so the next step was to determine the header length by using grep() to detect the end of the header. As more files were considered, it became desirable to infer data columns from the header, instead of specifying them manually. Other features were added as applications widened, and to avoid confusion the code was recast as a function that returned not just columnar data, but also other (meta-data) quantities, such as station number, sampling location, etc., which are sometimes present in CTD headers. With such additions, formal documentation became necessary, because even the author found it difficult to remember the features without examining the code. For oce, as perhaps for other packages, this was the time when the effort of packaging was seen to be worthwhile, in order not just to bind documentation and code together, but also to take advantage of the checks of code and documentation that are involved in R packaging and, importantly, to create a system that would benefit colleagues.

From the early stages, a version control system was used to track changes to the oce source code. Between 2007 and 2010, the subversion system was used, but then a switch was made to git. The code was originally hosted on a website on the author's desktop computer, but as the user base grew, it was moved to Google,[3] where it was called r-oce because the name oce had been taken. Then, in 2010, oce was moved to GitHub,[4] where it resides today, benefiting from the collaboration of additional authors and the advice and bug reports of users from around the world.

---

[3] code.google.com/p/r-oce.

[4] github.com/dankelley/oce.

The official version of oce is available on the CRAN[5] servers and may be installed with `install.packages()`. The Github website provides updates between official releases, and it is also used by those requesting new features, reporting bugs or otherwise helping with oce development.

It is worth noting that additions to oce are always based on the practical needs of the authors and their colleagues, never on some imagined needs. For example, support for CTD data was followed quickly by support for oceanographic sections, with the `section` class being added as a second child to the parent oce class. As the authors started working with acoustical instruments, support was added for acoustic-Doppler profilers (`adp`) and velocimeters (`adv`). This continued, one instrument at a time, with oce gradually growing to offer support for most instruments in common use today.

Generally, oce functions were developed to work on data in the authors' possession, often data under active study in a research program. An advantage of this (beyond satisfying individual research needs) was the early detection of coding errors or poor design. Through time, new features were increasingly based on requests from users, often as articulated on the development website[4]. By design, this scheme directs coding effort first and foremost to issues of high relevance to the oceanographic community.

At this point in the text, readers should be able to apply R to their own work, relying on oce to handle quirky data formats and produce diagrams in the oceanographic convention. However, as with any tool, there are dangers in forming habits based on success in early tests. For this reason, the remainder of this book addresses practical aspects to using R for oceanographic analysis, starting with a re-analysis of the data in some classic research articles, and then turning attention to more modern and technical issues.

---

[5] cran.r-project.org.

Chapter 4
Historical Examples

Abstract Data extracted from four highly influential oceanographic publications are used in step-by-step explanations of how to use R for modern-day analysis. The papers were chosen to represent the four sub-disciplines of oceanography, in hopes of providing an inherently interesting framework for a discussion of four key analytical tools.

LINEAR REGRESSION

> Alfred C. Redfield, 1934. On the proportions of organic derivations in sea water and their relation to the composition of plankton. James Johnstone Memorial Volume, p. 177–192. University Press of Liverpool.

DIFFERENTIAL EQUATIONS

> Gordon A. Riley, 1946. Factors controlling phytoplankton populations on Georges Bank. *Journal of Marine Research*, **6(1)**, p. 54–73.

TYPE II REGRESSION AND BOOTSTRAPPING

> J. Tuzo Wilson, 1963. Evidence from islands on the spreading of ocean floors. *Nature*, **197(4867)**, p. 536–538.

NONLINEAR REGRESSION

> Walter H. Munk, 1966. Abyssal recipes. *Deep-Sea Research*, **13**, p. 707–730.

4.1 Seawater Chemistry (Redfield 1934)

In the early twentieth century, significant efforts were made to measure the spatial and temporal patterns of seawater chemistry. Some amounted to exercises in mapping, while others were more focussed on deriving meaning from the relationships between measured quantities. A notable example of the latter was provided by

© Springer Science+Business Media, LLC, part of Springer Nature 2018
D. E. Kelley, *Oceanographic Analysis with R*,
https://doi.org/10.1007/978-1-4939-8844-0_4

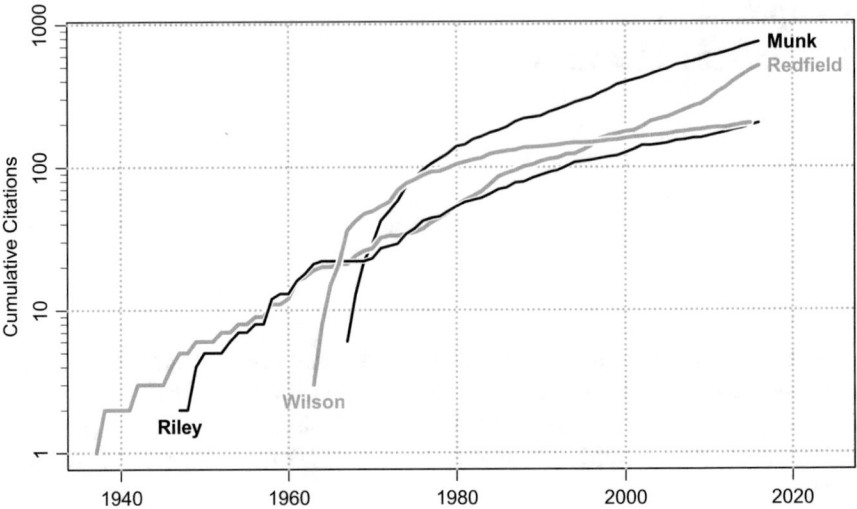

Fig. 4.1 History of citations of the papers listed above, according to a citation search using the Web of Science

Alfred Redfield in 1934, in his discussion of seawater chemistry in the context of plankton chemistry (Fig. 4.1 shows the history of citations of this paper, and of the others discussed in this chapter).

He reported that the concentration of Carbon, Nitrogen and Phosphorus in certain biologically active seawater compounds tend to be in constant ratio, C:N:P being 140:20:1 by mole (or 100:16.7:1.85 by mass). Importantly, he also showed that corresponding ratios within marine plankton were similar to those in seawater. Although the mechanisms behind the similarity were not entirely clear, the ramifications were so significant that Redfield (1934) came to be regarded as a founding paper in modern oceanography (Revelle 1995).

Redfield did not rely heavily on statistical analysis in his paper. This may reflect the times or his personal interests, but in either case, readers might find it helpful to see how to analyse his data in R. In the present treatment, the focus will be on the tools rather than the science, and concrete hints will be provided for graphical display, regression and hypothesis testing.

Data digitized from Figure 1 of Redfield (1934) are given in the `redfieldNP` dataset in the `ocedata` package, so the analysis starts with loading the data and using `str()` for an overview

```
data(redfieldNP, package="ocedata")
str(redfieldNP)
'data.frame':           119 obs. of  2 variables:
 $ PO4: num   0.0488 0.0694 0.075 0.0769 0.0919 ...
 $ NO3: num   0.63 0.607 1.349 1.535 0.63 ...
```

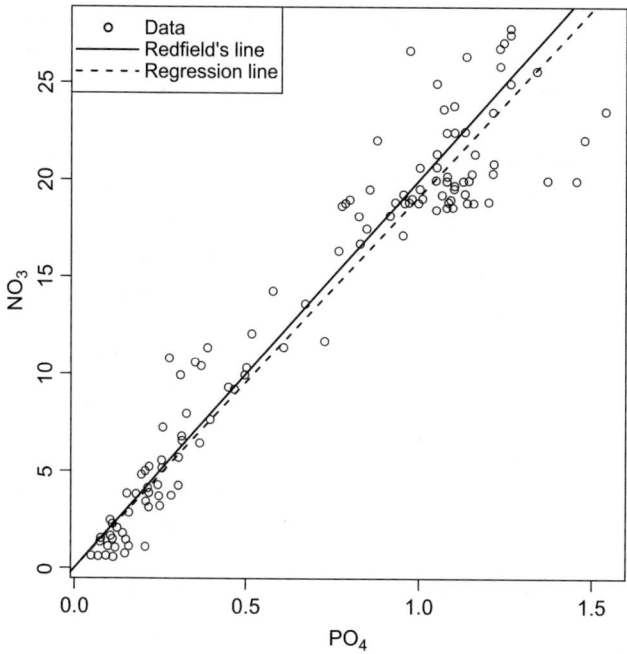

Fig. 4.2 Concentration of nitrate and phosphate (each in μmole/litre) as graphed in Figure 1 of Redfield (1934). The solid line has slope 1:20, as drawn by Redfield. The dashed line comes from a least-squares regression using `lm()`

The elements of this data frame may be accessed[1] with, e.g., `redfieldNP $PO4`, and a scatter plot (Fig. 4.2) produced with

```
plot(redfieldNP$PO4, redfieldNP$NO3,
     xlab=expression(PO[4]), ylab=expression(NO[3]))
```
while Redfield's N:P ratio of 1:20 is added with

```
abline(0, 20, lwd=2)
```

Now, we may go beyond Redfield's diagram. Since the relationship appears to be roughly linear, it is natural to fit a linear model relating nitrate to phosphate.[2] As discussed in Sect. 2.5.5, `lm()` is one way to fit such a model.

---

[1]The elements could be accessed as PO4 and not `redfieldNP$PO4`, etc., if preceded by an `attach(redfieldNP)` function call. However, some analysts, including the author, find that using `attach()` leads to confusion as to which variables are visible at any given time, and so they avoid the function and its pair, `detach()`.

[2]Recognizing that there are errors in both nitrate and phosphate, one might also try a type-II regression, a matter to be discussed more in Sect. 4.3.

Based on Fig. 4.2 plus a desire to explore the concept of Redfield ratio, one might seek a linear relationship between NO_3 and PO_4, with intercept set to zero. The intercept is removed by using -1 in the regression formula[3]

```
m <- lm(NO3 ~ PO4 - 1, data=redfieldNP)
```

where the regression result has been saved in m, so that it can be used later. The first use is in adding a regression line to Fig. 4.2

```
abline(m, lty="dashed", lwd=2)
```

where abline() detects that m was created by lm(), and so it extracts the slope and intercept automatically.[4] Finally, a legend is created with

```
legend("topleft", pch=c(1, NA, NA),
        lty=c(NA, "solid", "dashed"), lwd=2, seg.len=4,
        legend=c("Data", "Redfield's line",
        "Regression line"))
```

An overview of the model fit is given by summary(), a generic function (Sect. 2.3.11.6) that in turn calls summary.lm(); the full output is shown below.

```
summary(m)
Call:
lm(formula = NO3 ~ PO4 - 1, data = redfieldNP)

Residuals:
    Min      1Q  Median      3Q     Max
-7.9064 -1.1923  0.0619  1.2407  8.0237

Coefficients:
     Estimate Std. Error t value Pr(>|t|)
PO4   19.1560     0.2726   70.28   <2e-16 ***
---
Signif. codes:
0 '***' 0.001 '**' 0.01 '*' 0.05 '.' 0.1 ' ' 1

Residual standard error: 2.45 on 118 degrees of freedom
Multiple R-squared:  0.9767,   Adjusted R-squared:  0.9765
F-statistic:  4939 on 1 and 118 DF,  p-value: < 2.2e-16
```

The low overall p value suggests that the regression is statistically significant, and the high adjusted R^2 value indicates that the regression accounts for most of the nitrate variability. The residual standard error reported by summary() also tends to be a focus of interest, in cases where replicates or other schemes provide information on the random measurement errors.

---

[3]Note the use of the data argument to lm(), which exposes the elements of redfieldNP, so that, e.g., NO3 can be written instead of redfieldNP$NO3 within the function call.

[4]The line could also be drawn with lines(), with the output from predict(); see a nonlinear example in Sect. 2.5.5.2.

The inferred slope is lower than Redfield's value of 20, at least as indicated by the 95% confidence interval

```
confint(m)
            2.5 %    97.5 %
PO4 18.61619 19.69572
```

The next step is to consider the concentrations of elements in plankton, which Redfield (1934) presented in weight-ratio form in his Table II. These values are in the dataset redfieldPlankton in the ocedata package.

```
data(redfieldPlankton, package="ocedata")
```

Following Redfield , we may examine the averages (using na.rm in the second case because it contains some NA values)

```
mean(redfieldPlankton$Nitrogen)
[1] 15.44545
```

```
mean(redfieldPlankton$Phosphorus, na.rm=TRUE)
[1] 1.88
```

The values within redfieldPlankton are masses expressed as a percent of Carbon mass. Redfield compared the averages with seawater values 16.7 and 1.85 (in weight terms) in his Table II. It is perhaps revealing that Redfield did little more than present these values, writing that the values "are not greatly different." Such laudable brevity notwithstanding, the present purpose is to illustrate the use of R, so further steps may be warranted.

A common approach would be to use t.test() to perform a *t* test to compare the data with Redfield's stated values. The seawater value 16.7 could be used as a value for comparison:

```
t.test(redfieldPlankton$Nitrogen, mu=16.7)
            One Sample t-test

data:   redfieldPlankton$Nitrogen
t = -0.86503, df = 10, p-value = 0.4073
alternative hypothesis: true mean is not equal to 16.7
95\% confidence interval:
 12.21401 18.67690
sample estimates:
mean of x
 15.44545
```

With such a large *p* value, it is difficult to argue that the plankton ratio is different from the seawater value. Another indication is the fact that the mu value lies within the confidence interval of 12.2 to 18.7.

As stated, Redfield did not pursue the agreement in any great depth. Instead, he looked to the future, ending his paper by writing

Whatever its explanation, the correspondence between the quantities of biologically available nitrogen and phosphorus in the sea and the proportions in which they are utilized by the plankton is a phenomenon of the greatest interest.

These words still ring true all these decades later, testifying not just to the significance of Redfield's ideas, but also to his ability to explain them. Further to this second point, readers who study this man will soon find that he had a flair for communication. Indeed, according to Revelle (1995), Redfield was fond of saying

> Life in the sea cannot be understood without understanding the sea itself.

and one would be hard-pressed to find a better statement of the gist of modern biological oceanography.

Exercise 4.1 Use loess() to fit localized polynomial models (Sect. 2.6.6) of PO_4 as a function of NO_3, and vice versa. (See page 211 for a solution.)

Exercise 4.2 Alter the lm() call for the Redfield ratio fit, to test whether the slope might be 20. (See page 211 for a solution.)

Exercise 4.3 Calculate the slope in Fig. 4.2 using ridge, robust and resistant regression, using functions lm.ridge(), rlm() and lqs() from the MASS package, respectively. (See page 212 for a solution.)

4.2 Ecosystem Modelling (Riley 1946)

One of many contributions Gordon Riley made to modern biological oceanographers was to illustrate the utility of using differential equations to track temporal variations of interconnected elements. A prime example of this is found in his 1946 analysis of phytoplankton populations on Georges Bank, which pioneered a setup that is still used to this day, with some variations and extensions.

Before getting to the methodology of Riley's differential equation solution, it may be instructive to quote the first words of his 1946 paper:

> A complex field such as oceanography tends to be subject to opposite approaches. The first is the descriptive, in which several quantities are measured simultaneously and their interrelationships derived by some sort of statistical method. The other approach is the synthetic one, in which reasonable although perhaps over-simplified assumptions are laid down, these serving as a basis for mathematical derivation of relationships ... in many cases there is no chance for mutual profit because the two approaches have no common ground. Until such contact has been established no branch of oceanography can quite be said to have come of age.

This was an important message for a field that was at that time finding its way in the scientific world. Indeed, it is no less important today; one might argue that Riley's "mutual benefit" is at the heart of effective modern oceanographic work.

In the first part of his paper, Riley presented a descriptive treatment that would have been familiar to fellow marine biologists of the day. Then, in a move that might have surprised many of his readers, he turned his attention to a differential equation. This equation related temporal changes in P, the depth-integrated phytoplankton concentration (in grams carbon per square metre of surface), to gains by photosynthesis and losses by respiration and grazing

$$\frac{dP}{dt} = (P_h - R - G)\, P \tag{4.1}$$

where t is time, and the rate factors P_h, R and G describe the effects of photosynthesis, respiration and grazing.

In dealing with phytoplankton *en mass*, in ignoring such effects as advection and diffusion, and in using linear terms on the right-hand side of (4.1), Riley was providing a good example of what he called "perhaps over-simplified assumptions" yielding a plausible model that was easy to solve.

As an important further simplification, Riley took the rate factors to be constant in prescribed time intervals. This turned (4.1) into a set of linear differential equations with constant coefficients, from which he could construct a piecewise exponential solution. This made for a simple calculation, entirely appropriate for his purpose. However, variable rate constants are easy to handle with today's numerical integration tools, and so a good way to illustrate modern procedures for dealing with differential equations is by extending Riley's setup to allow the rate constants to vary continuously with time.

The dataset `riley` holds material that can be used to produce Fig. 4.3, which compares Riley's solution with one developed in R. The first step is to load a digitized approximation of Riley's data

```
data(riley, package="ocedata")
```

and plot his observations along with his solution (digitized from his paper)

```
plot(riley$fig21points$day, riley$fig21points$P, pch=20,
     xlab="Day", ylab="Phytoplankton")
lines(riley$fig21curve$day, riley$fig21curve$P)
```

The present numerical integration will be done with the `deSolve` package, although other packages also offer integrators (Soetaert et al. 2010).

```
library(deSolve)
```

The function `lsoda()` integrates initial value problems of the form $dx/dt = f$, to find x, where t is an independent variable such as time, and f is a forcing term.

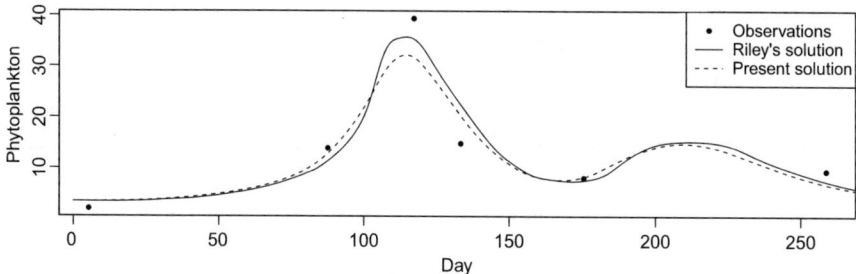

Fig. 4.3 Variation of depth-integrated phytoplankton concentration on Georges Bank. The dots indicate data shown by Riley (1946) in his Figure 21, and the solid curve was digitized from the solution he indicated on that figure. The dashed line is the present solution for piecewise-linear forcing, using `lsoda()` to perform the integration

Readers familiar with numerical integration should note that lsoda() can handle a wide variety of problems, because it monitors the solution and adjusts the algorithm automatically to balance accuracy and efficiency, e.g. switching between stiff and non-stiff solvers as needed.

A function must be provided to lsoda() to describe the right-hand side of (4.1). The variation of P_h, R and G with time is given at roughly 2-week intervals in Riley's appendix, and these values are made available in riley$DEparameters. Lacking data on variations within the 15-day intervals, it is necessary to make some further assumptions. A simple one is piecewise-linear variation, for which approxfun() is useful[5]

```
funPh<-approxfun(riley$DEparameters$day,
riley $DEparameters$Ph)
funR <- approxfun(riley$DEparameters$day,
riley $DEparameters$R)
funG <- approxfun(riley$DEparameters$day,
riley $DEparameters$G)
```

It remains to define a function to be integrated by lsoda(). This must take the independent variable as its first argument, the dependent variable as its second, and a list of parameters as its third, and it must return a list containing the derivative. Therefore, (4.1) may be handled with

```
f <- function(t, P, parameters)
    list(P*(funPh(t) - funR(t) - funG(t)))
```

The first argument to lsoda() is the initial condition, while the second is a vector of reported values of the independent variable, the third is the function and the fourth is a list of parameters. In the present case, a numerical solution to (4.1), reported on a daily interval and with initial condition $P = 3.4$, is

```
solution <- lsoda(3.4, 1:365, f, NULL)
```

where NULL indicates that f() makes no use of parameters.

The return value is a matrix with first column holding the prescribed values of the independent variable, and succeeding column(s) representing the dependent variable(s), so

```
lines(solution[,1], solution[,2], lty="dashed")
```

adds numerical solution to Fig. 4.3. The plot indicates reasonable agreement between the two methods.[6]

Given the effort this sort of calculation would have entailed in the 1940s, Riley's simpler approach was both practical and effective. Besides, Riley was not proposing his work as a deep solution, contextualizing it as follows.

> While these methods are obviously crude at the present time and need to be developed further … it does not seem too much to hope that they will eventually solve some of the problems of seasonal and regional variations that puzzle marine biologists today.

---

[5]Readers might like to try other approaches, e.g. fitting a spline function (Sect. 2.6.7).

[6]A similar finding is reported by Anderson and Gentleman (2012).

Exercise 4.4 Use `lsoda()` to solve the NPZ equations as expressed in Chapter 4 of Sarmiento and Gruber (2006)

$$\frac{dN}{dt} = P\left(-V_{max}\frac{N}{K_N + N} + \mu_P\,\lambda_P\right) + Z\,\mu_Z\left[(1-\gamma_Z)\,g\,\frac{P}{K_P} + \lambda_Z\right]$$

$$\frac{dP}{dt} = P\left(V_{max}\frac{N}{K_N + N} - \lambda_P - \frac{g\,Z}{K_P}\right) \qquad (4.2)$$

$$\frac{dZ}{dt} = Z\left(\gamma_Z\,g\,\frac{P}{K_P} - \lambda_Z\right)$$

with $V_{max} = 1.4\,\mathrm{d}^{-1}$, $K_N = 0.1\,\mathrm{mmol/m^3}$, $\mu_P = 1$, $\lambda_P = 0.05\,\mathrm{d}^{-1}$, $\mu_Z = 1$, $\gamma_Z = 0.4$, $g = 1.4\,\mathrm{d}^{-1}$, $K_P = 2.8\,\mathrm{mmol/m^3}$, and $\lambda_Z = 0.12\,\mathrm{d}^{-1}$. Use initial conditions $N = 10$, $P = 3$ and $Z = 2$, and plot the results over a month. (See page 212 for a solution.)

Exercise 4.5 Use `lsoda()` to develop a numerical solution to the wave equation $d^2\eta/dt^2 + \omega^2\eta = 0$ with frequency $\omega = 1\,\mathrm{s}^{-1}$ during $0 \le t \le 2\pi$, with initial condition $\eta = 0$ and $d\eta/dt = 1$ at $t = 0$. (See page 213 for a solution.)

4.3 Plate Tectonics (Wilson 1963)

With the mid-twentieth century discovery of banded magnetic fields near the ocean bottom, the old idea of continental drift gained new life, and evidence was woven quickly into the plate tectonics theory (Wilson 1962; Vine and Matthews 1963; Backus 1964). An important thread was J. Tuzo Wilson's study of ocean island age, which he motivated by writing (Wilson 1963)

> If the Earth has been rigid, the history of the ocean basins should parallel that of the continents and large parts of the ocean floor should be old; but, if continents have moved, those parts of the floors exposed by the motion should not be older than the time of drifting. Explorations of the ocean floors may settle this matter, but in the meantime information has been sought from the literature published about ocean islands.

Wilson's Figure 2 provided a clear illustration that there was merit in further work along these lines. That diagram can be echoed as Fig. 4.4 with

```
data(wilson, package="ocedata")
plot(wilson$Age, wilson$Distance,
     xlab="Age [My]", ylab="Distance [km]",
     xlim=c(125, 0), ylim=c(0, 4000))
abline(0, 4000/125)
```

where `abline()` adds Wilson's guiding line. While this line certainly served the purpose, we can use a search for alternatives as a motivation to explore regression tools in R. For example, a least-squares regression line passing through the origin is developed with

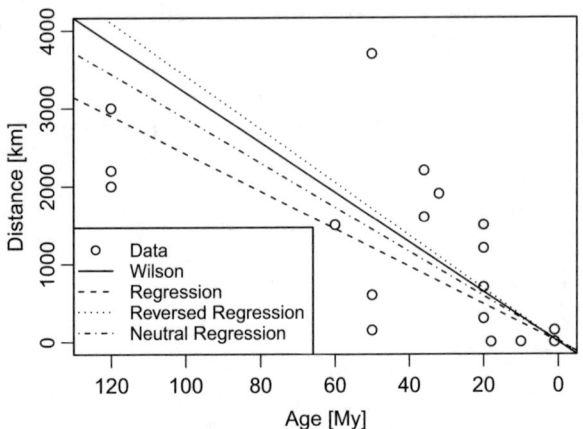

Fig. 4.4 Re-analysis of seafloor-spreading data from Table 1 of Wilson (1963)

```
m0 <- lm(Distance ~ Age - 1, data=wilson)
abline(m0, lty=2)
```
and one assuming errors in `Age` rather than `Distance` with
```
mr <- lm(Age ~ Distance - 1, data=wilson)
abline(a=0, b=1/coef(mr), lty=3)
```
A study of the differences might motivate a neutral regression (Sect. 2.5.5), perhaps using the `smatr` package, with
```
library(smatr)
mn <- sma(Distance ~ Age, data=wilson)
abline(coef(mn), lty=4)
```
showing a result within the expected range of the others.

Wilson's study was exploratory, and one might wonder how its results might have varied with alternative data. One way to address this is to use bootstrapping, a procedure in which a calculation is performed repeatedly on random subsets of the data (Efron and Gong 1983; Efron and Tibshirani 1998). This can provide interesting insights for oceanographic measurements, which often violate the assumptions of traditional statistical tests.

It is easy to code a bootstrapping procedure in R, but it is better to use the `boot` package, in order to get standardized confidence intervals and graphical displays. The first argument of this function is a data matrix
```
d <- cbind(wilson$Age, wilson$Distance)
```
and the second is a function yielding a quantity of interest, based on the matrix and an indexing vector. For example, it could be a regression slope

```
s   <- function(d, i) coef(sma(d[i,2]~d[i,1]-1))
[["slope"]]
```

and, with this, a bootstrap with 100 replicates can be done with

```
library(boot)
b <- boot(d, s, R=100)
```

Confidence limits on the results are calculated with

```
boot.ci(b)
BOOTSTRAP CONFIDENCE INTERVAL CALCULATIONS
Based on 100 bootstrap replicates

CALL :
boot.ci(boot.out = b)

Intervals :
Level        Normal              Basic
95%    (19.88, 37.10 )    (20.23, 36.50 )

Level       Percentile             BCa
95%    (20.95, 37.22 )    (20.19, 36.61 )
Calculations and Intervals on Original Scale
Some basic intervals may be unstable
Some percentile intervals may be unstable
Some BCa intervals may be unstable
```

and it is worth noting that Wilson's estimate of 35 km per million years is within them.

Exercise 4.6 Explore type II regression with the lmodel2 package. (See page 214 for a solution.)

4.4 Ocean Mixing (Munk 1966)

Many of the foundations of the present-day understanding of ocean dynamics were established in the middle of the twentieth century, by just a few individuals. One of them was Walter Munk, who contributed greatly to a dizzying list of topics in physical oceanography and in geophysics, generally.

An important example is his study of the thermocline in Pacific central waters (Munk 1966). This paper motivated decades of work on the measurement and theory of ocean mixing (see Gregg 1991, for early historical notes) and connections to large-scale ocean and climate systems (see Melet et al. 2016, for a recent treatment). Although Munk wrote

> The model is not new: it was used by Wyrtki (1962) in a discussion of the oxygen minimum and in various forms goes back to oceanographic antiquity

Fig. 4.5 Definition sketch
for a crude thermocline
model.

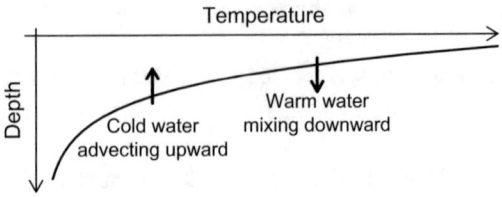

the model is called "Munk's model" by almost everyone in the field.

The idea is sketched in Fig. 4.5. Heat supplied at the surface is mixed downward, but the resulting warming of mid-depth waters is countered by the upwelling of colder waters. A steady one-dimensional model[7] of the system is

$$0 = -w\frac{\partial\theta}{\partial z} + \frac{\partial}{\partial z}\left(K_V\frac{\partial\theta}{\partial z}\right) \tag{4.3}$$

where w is the upwelling velocity, θ is the potential temperature, z is the vertical coordinate increasing upwards to $z = 0$ at the surface, and K_V is a so-called "turbulent" or "eddy" diffusivity, analogous to molecular diffusivity.

If w and K_V can be approximated as constants in a given depth zone, then (4.3) becomes an ordinary differential equation with constant coefficients. With surface boundary condition $\theta = \theta_S$ and far-field boundary condition $\theta \to \theta_D$ as $z \to -\infty$, the solution is

$$\theta = \theta_D + (\theta_S - \theta_D)\exp(z/h) \tag{4.4}$$

where $h = K_V/w$ is a thermocline scale thickness.

If w can be constrained by some means, then inferring h from a potential temperature profile will reveal the value of K_V, which is an important parameter for more general models.

Figure 4.6 contains an excerpt of Munk's Figure 1, indicating his inference of $h \approx 0.77$ km (the reciprocal of the label $w/\kappa = 1.3$). Data digitized from Munk's diagram are provided by the ocedata package, making it easy to see how Munk's visual technique compares with nonlinear regression in R.

The first step is to extract the data

```
data(munk, package="ocedata")
theta <- munk$temperature # file stores pot. temp.
z <- -1000 * munk$depth # convert depth [km] to z [m]
```

and plot as in Fig. 4.6, with

```
plot(theta, z, type="p", xlim=c(0.5, 4.0),
ylim=c(-4000, 0),
    xaxs="i", yaxs="i",xlab="Potential temperature")
```

---

[7]The 1D simplification calls to mind Gordon Riley's advice, quoted in Sect. 4.2, but one should note that Munk explored more complicated models in his paper.

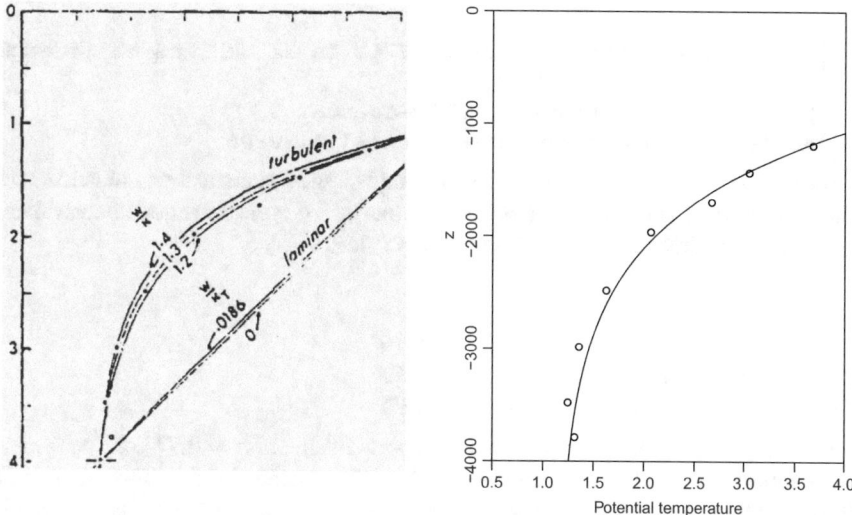

Fig. 4.6 Left: excerpt from Figure 1 of Munk (1966). The vertical axis is depth in kilometres and the horizontal axis is potential temperature, ranging from 0.5°C to 4°C. The curves are as (4.4) for a variety of values of w/K_V, including nearly flat curves for molecular diffusivity. Right: data digitized from Munk's figure, along with the prediction of a regression model.

where limits are chosen to match the geometry of Munk's figure.

A nonlinear regression to the theoretical form may be carried out with[8]

```
m <- nls(theta ~ thetad + (thetas-thetad)*exp(z/h),
         start=list(thetas=5, thetad=1, h=1000))
```

The predictions of this regression model are shown in Fig. 4.6 with

```
zz <- seq(from=-4000, to=0, by=100)
TT <- predict(m, newdata=data.frame(z=zz))
lines(TT, zz)
```

and the results may be summarized with

```
summary(m)
```

```
Formula: theta ~ thetad + (thetas - thetad) * exp(z/h)

Parameters:
          Estimate Std. Error t value Pr(>|t|)
thetas    11.23138    0.59128   19.00 1.05e-14 ***
thetad     1.16583    0.02019   57.76  < 2e-16 ***
h        844.65390   35.46673   23.82  < 2e-16 ***
---
Signif. codes:
0 '***' 0.001 '**' 0.01 '*' 0.05 '.' 0.1 ' ' 1
```

---

[8]Sensible starting parameter values are required for convergence; see Sect. 2.5.5.2.

```
Residual standard error: 0.07741 on 21 degrees of freedom

Number of iterations to convergence: 8
Achieved convergence tolerance: 1.815e-06
```

Note that θ_S, θ_D and h each have p values that suggest statistical significance, and that the residual standard error is comparable to the typical scatter in such data. Computing confidence intervals on the fitted values

```
confint (m)
                 2.5%         97.5%
thetas    10.132192    12.515536
thetad     1.123981     1.207263
h        777.102870   919.107091
```

reveals slight disagreement with Munk's reported value of $h = 0.77$ km.

A follow-up paper by Munk and Wunsch (1998) extended some of the ideas, and it too provides a useful key for a literature search on ocean mixing. Such a search will reveal that direct measurements of ocean turbulence have tended to contradict a naive interpretation of Munk's K_V value, lending weight to the pithy ending of his prescient 1966 paper:

> Until the processes giving rise to diffusion and advection are understood, the resulting differential equations governing the interior distribution, and their solutions, must remain what they have been for so long: a set of recipes.

Exercise 4.7 Use try() to skip past errors in nls(), so that a bootstrap estimate of h can be done. (See page 215 for a solution.)

Exercise 4.8 Specify gradients to nls() to fit the munk data. (See page 116 for a solution.)

4.5 Concluding Remarks

It is a sad fact that classic papers of the type selected for this chapter are sometimes cited by those who have not read them in much detail. For example, some of the citations to Munk's paper display a peculiar error that suggests that the citing authors did not look at the original paper, but instead derived bibliographic information from a published work in which the original paper had been cited incorrectly. Behaviour of this sort is unfortunate, not just in terms of missed learning opportunities, but also in terms of missed chances for motivation. This explains why this chapter is framed around classic papers, and why there are so many quotations from them. There can be great joy in rediscovering the birth of a field.

Admittedly, the treatment here is somewhat strained, stemming from a desire to (a) explain some useful R methods in an interesting context, and (b) distribute the material across four sections that touch on the four main sub-disciplines of oceanography.

It should be noted that no attempt has been made here to match the R topic to the thrust of the cited papers, e.g. Munk was unlikely to have been thinking of nonlinear regression when he cooked his abyssal recipes! And it should go without saying that there is no intention to suggest that methods such as those outlined here should have been employed by the original authors. Indeed, readers who explore further will notice that the authors were quite conscious that their work was limited. As it turns out, though, the R approaches have mainly confirmed the results obtained by the original investigators, and this will not surprise those familiar with the literature, who know that these authors cast long shadows for good reason. Deep thinkers are often careful workers, achieving trustworthy results with simple techniques.

In the works considered here, graphical methods were often the most appropriate tool. This is still true for exploratory work today, although the need to examine larger datasets motivates the incorporation of statistical and data-reduction techniques as well. These are the "descriptive" and "synthetic" approaches of Gordon Riley, and the common integration of the two indicates that, in Riley's words quoted on page 108, oceanography has "come of age." The next chapter illustrates this descriptive-synthetic integration in a diverse series of short examples.

Chapter 5
Practical Operating Procedures

Abstract Oceanographers are commonly called upon to analyse datasets that are marred by instrument malfunction and loss, or the inability to resolve features that are of an unexpected form. Repeating experiments is not the option it is in bench-based science, so analysts must become skilled at developing ad hoc procedures to wring useful information out of whatever data can be acquired. This is why standard operating procedures are less common in oceanography than what might be called "practical operating procedures." R is an ideal tool for this way of working, given its inherent suitability for interactive analysis and its provision of a vast array of libraries for specialized techniques.

5.1 Introduction

Some oceanographic work takes place in laboratories, where extraneous factors can be reduced, instruments can be maintained and experiments can be repeated. Using standard operating procedures for data analysis can increase productivity and reliability in such situations. However, things are different at sea, where phenomena are too interconnected to permit controlled experiments, instruments are routinely damaged or lost, and high environmental variability makes every repeated experiment unique in some way. Coping with unplanned aspects of fieldwork can require so much ad hoc adjustment that there is little hope of true standardization. Instead, field oceanographers tend to employ what might be called practical operating procedures, adjusting methods to get the most from available data.

Oceanographers tend to work with many different types of data in any given project. Using a separate analysis tool for each data type can offer benefits of specialization, but only at the cost of time spent learning to use individual tools. Also, the tools tend to be limited, because of the burden of adding commonly required capabilities (e.g. the equation of state) to each. It makes sense to use integrated tools, and the main message of this book is that R can take a place in the centre of such tools. Certainly, it has good procedures for the analysis of clean datasets, but there are several competitors in this category. Where R really shines

© Springer Science+Business Media, LLC, part of Springer Nature 2018
D. E. Kelley, *Oceanographic Analysis with R*,
https://doi.org/10.1007/978-1-4939-8844-0_5

is with the problematic datasets that oceanographers often face. It is this balanced applicability that makes R such a useful tool for general oceanographic analysis, as will be illustrated in the pages that follow.

The needs of bench-based oceanographers are generic enough to be covered well by most R textbooks, and so the main focus here is on the analysis of field observations. This topic has been covered previously in both specific and general ways. As examples of the former, Mamayev et al. (1991) outline the basics of processing CTD station data, and IOC et al. (2010) detail the care that must be taken with the equation of state. More general treatments of physical oceanography are provided by a series of texts, including the expansive treatment by Talley et al. (2011). There are also books that survey methods of analysis, e.g. Emery and Thomson (2001) and Wunsch (2015) deal with physical oceanography and Glover et al. (2011) add in a modelling view, while numerical ecology is treated by Legendre and Legendre (1998) and (in an R context) by Borcard et al. (2011). Wide reading also pays off for oceanographers, e.g. many methods for processing atmospheric data explained by Daley (1991) apply to the ocean as well. The general literature is also important, and it seems fair to say that most oceanographers' bookshelves hold monographs on time-series analysis, spatial analysis, neural networks, etc., alongside books about the sea.

5.2 Hydrography

5.2.1 Seawater Calculation

As explained in Chap. 3, the oce package provides many functions that work with seawater properties. Their names begin with sw, and so their documentation may be found with

```
help.search("sw", package="oce")
```

The documentation is extensive and tied to the peer-reviewed literature, so a summary sketch should suffice here.

The sw functions that can accept oce objects as the first argument try to look within the object to find all requisite information, e.g. in

```
data(ctd, package="oce")
head(swRho(ctd, eos="gsw"), 3)
[1] 1022.23316 1022.23254 1022.23460
```

swRho() has used the longitude and latitude values stored within ctd for the GSW calculation of seawater density (see Chap. 3 and Appendix D).

The accessors for the oce objects that hold hydrographic data are tailored somewhat to the object type, e.g.

```
data(section)
SAv <- section[["SA"]]
```

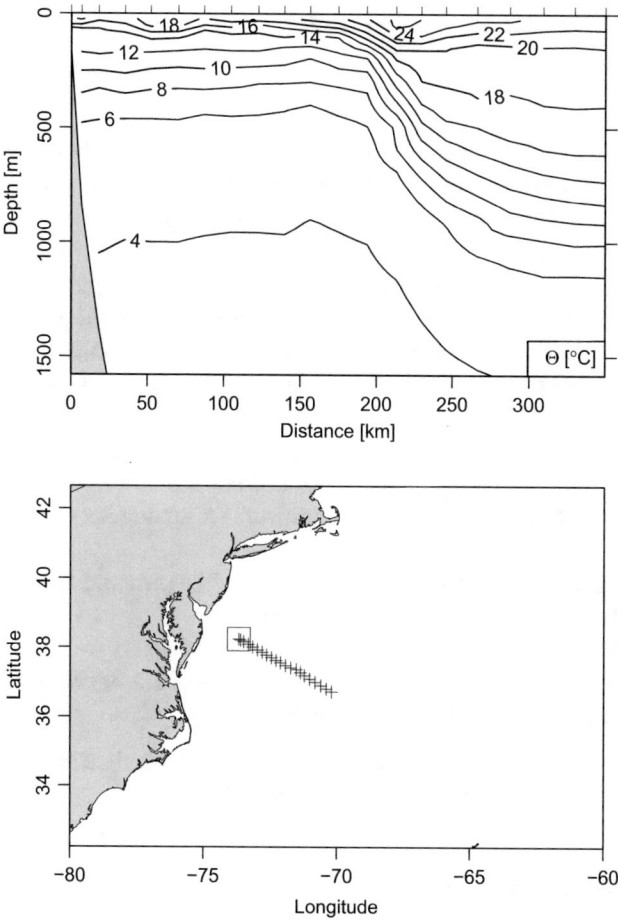

Fig. 5.1 Top: Conservative Temperature section across the Gulf Stream; bottom: station locations, with a box indicating the station from which distance is measured in the section. (See Page 129)

returns a vector holding all the Absolute Salinity values in the section, while

```
SA1 <- section[["SA", "byStation"]]
```

returns a list containing values split by station. (These station locations are shown in the lower panel of Fig. 5.1, a diagram explained further in Section 5.2.2.8.)

Although many oce functions have an eos argument, it is also common to use gsw functions (Kelley et al. 2017) directly, e.g.

```
head(gsw_rho(ctd[["SA"]], ctd[["CT"]],
    ctd[["pressure"]]), 3)
```

yields results matching the swRho() results shown above, and

```
cabbel <- gsw_cabbeling(ctd[["SA"]],
    ctd[["CT"]], ctd[["p"]])
```

calculates a cabbeling parameter that is not provided by oce.

5.2.2   Profile and Section Analysis

5.2.2.1   Reading CTD Data

The `oce` package provides input functions for a wide variety of oceanographic instruments, so readers will not need to do any special coding unless they have an uncommon instrument. Generally, one may write, e.g.,

```
d <- read.oce("sta01.cnv")
```

which detects the type of file and then calls a specialized input function (e.g. the above will call `read.ctd.sbe()` to read a Seabird CTD file). The result is an object with a `class` reflecting the data type. For most data types, data read by some other means can also be transformed into such an object, e.g.

```
ctd <- as.ctd(S, T, p)
```

will create a CTD object given salinity, temperature, and pressure; additional arguments permit the specification of longitude and latitude.

It is common to store CTD files for a project in a subdirectory, and e.g.

```
pdf("ctd.pdf")
for (file in list.files(path=".", pattern=".cnv$"))
    plot(ctdTrim(read.oce(file)))
dev.off() # close multipage PDF file
```

will plot them all as pages in a PDF file. This can be a convenient first step in processing, especially if the file names order sensibly.

Exercise 5.1 Use `sub()` to create a series of PDF files with names that map to the names of data files. (See page 217 for a solution.)

5.2.2.2   Editing CTD Data

It is possible to edit (alter) the CTD data using, e.g.,

```
data(ctd, package="oce")
ctd[["temperature"]]  <- ctd[["temperature"]] + 0.1
```

to add a 0.001°C, but it is much better to write

```
ctd <- oce.edit(ctd, "temperature",
ctd[["temperature"]]+0.1)
```

because `oce.edit()` stores a record in the object's processing log, indicating that the original value was changed. This function can also be used to alter metadata, e.g.

```
ctd <- oce.edit(ctd, "waterDepth", 100)
```

makes the change in the `metadata` slot for this particular object, because that's where `waterDepth` is stored for CTD objects. Readers who try the above will see that `summary(ctd)` lists the two recent edits, as well as one that was required to correct a date problem in the original `.cnv` file used to create the `ctd` dataset provided by `oce`.

5.2.2.3 Plotting CTD Data

A generic function in the oce package provides a good starting point for examination of CTD data, e.g.

```
data(ctd, package="oce")
plot(ctd)
```

produces Fig. 3.1 on page 94. Additional arguments enable customization, with further refinement being available with plotTS() and plotProfile().

5.2.2.4 Trimming CTD Data

With a glance at Fig. 3.1, experienced oceanographers will suspect that a measured profile must have been trimmed to just the "downcast" phase, during which the CTD was lowered at a nearly constant speed to its deepest point. Such trimming is a common first step in the analysis of CTD profiles, and so the oce package provides functions to help in the work. The ctdRaw dataset provided with the package is useful for illustration.

It is convenient to start CTD analysis with a summary, and

```
data(ctdRaw, package="oce")
summary(ctdRaw)
```

produces results that include

	Min.	Mean	Max.	Dim.	Original Name
...					
temperature [°C, ITS-90]	2.3226	8.0027	98.952	773	t068
salinity [PSS-78]	0.3276	28.504	99	773	sal00

The salinity minimum is odd since only a small river runs into Halifax Harbour, where the profile was done, and the maximum is definitely suspicious, given its value and the fact that bad data are often set to 99 in oceanographic data files. The temperature maximum is also suspicious, being the ITS-90 equivalent of 99°C in the IPTS-68 scale used by this particular CTD.[1]

Of course, maxima and minima identify only single points, but larger sets are revealed with graphical and statistical tools, e.g.

```
plotScan(ctdRaw, type="o")
```

produces Fig. 5.2, suggesting a single spurious pressure. More statistically based graphical tests include

```
p <- ctdRaw[["pressure"]]
stem(p)
boxplot(p)
hist(p)
```

---

[1] When oce summarizes or otherwise reports data on CTD data, it converts to modern units and scales.

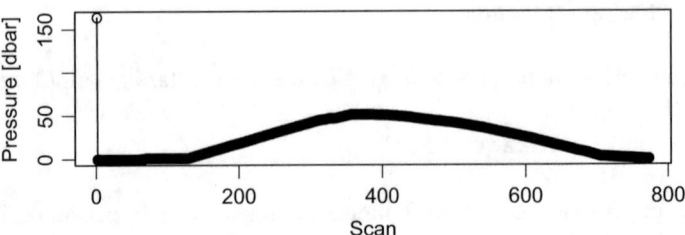

Fig. 5.2 CTD scan plot, revealing an initially unrealistic pressure value, followed by an equilibrating phase, a downcast, and then an upcast

and numerical tests include the following, all of which reveal that the first pressure is the only worrisome value.

```
which(abs(diff(p)) > 1)
which(abs(p - smooth(p)) > 0.5)
which(abs(p - mean(p)) > 2*sd(p))
```

Readers are likely to have other ideas for dealing with this or other problematic datasets, by combining simple R actions. This may involve looking at multiple variables at once, and considering dynamical considerations such as the expected tendency of density to increase with depth. (For more on such matters, see Sects. 5.9.2.1 and 5.9.4.6.)

Once suspicious data are identified, it is common to use quality-control flags to indicate the problem, and the `oce` function `handleFlags()` can help with dealing with these flags, as will be illustrated repeatedly in this chapter.

In addition to ignoring spurious data, oceanographers commonly also ignore the equilibration phase of CTD casts (in which the instrument is held below the surface to equilibrate with the seawater), and also the upcast that brings the package back up to the surface from deep water. The `oce` function `ctdTrim()` can be used for this, although for detailed work it may make sense to do things semi-manually, as in Exercise 5.4.

Exercise 5.2 Explore the sensitivity of buoyancy frequency, calculated with `swN2()`, to the argument `df`. (See page 217 for a solution.)

Exercise 5.3 Plot salinity and temperature profiles for the `ctd` dataset within 3 dbar of the pycnocline centre. (See page 218 for a solution.)

Exercise 5.4 Use `ctdTrim()` and `plotScan()` together, to trim `ctdRaw` to just the downcast portion. (See page 219 for a solution.)

5.2.2.5   Smoothing and Decimating CTD Data

CTDs are typically set up to record data on a fixed time interval, but it is common to store data in a fixed space interval, after smoothing the data in some way. A popular

form of smoothing is boxcar averaging, also known as bin averaging. Software
supplied by instrument manufacturers can do this, but it can also be done in R.
The `ctdRaw` dataset, trimmed to downcast,

```
data(ctdRaw, package="oce")
ctd2 <- ctdTrim(ctdRaw, method="sbe")
```

can be used for demonstration. The first step is to choose a pressure increment Δp
between averaging bins, and

```
summary(diff(ctd2[['pressure']]))
    Min. 1st Qu.  Median    Mean 3rd Qu.    Max.
 -0.1430  0.0470  0.1900  0.1627  0.2390  0.6200
```

might suggest using $\Delta p = 1$ dbar, which will include about 5 points per sample,
perhaps sufficient to average across noise.

It so happens that 1 dbar is the default used by `ctdDecimate()`, so

```
data(ctdRaw, package="oce")
ctd2 <- ctdTrim(ctdRaw, method="sbe")
ctd3 <- ctdDecimate(ctd2)
```

handles the task at hand. Figure 5.3, created with

```
par(mfrow=c(1,2))
plot(ctd2, which="CT", plim=c(15, 0))
lines(ctd3[["CT"]], ctd3[["pressure"]],
lwd=3, col="gray")
plot(ctd2, which="SA", plim=c(15, 0))
lines(ctd3[["SA"]], ctd3[["pressure"]],
lwd=3, col="gray")
```

shows the results near the top of the water column, for Conservative Temperature
and Absolute Salinity. Note the reduction in point-by-point variation.

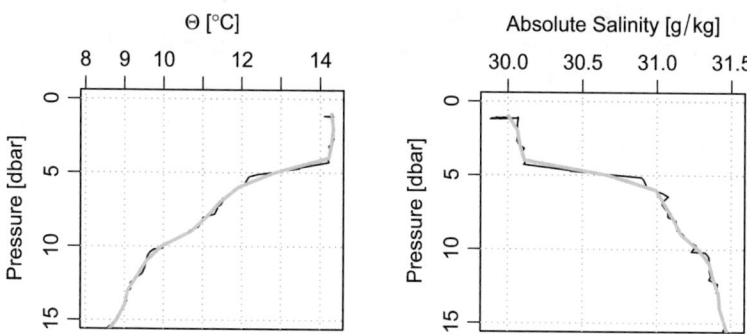

Fig. 5.3 Demonstration of CTD average-decimate scheme, with black line segments for data and
gray ones for the decimated result

5.2.2.6   Mixed-Layer Detection

Various factors cause turbulence near the surface of the ocean. The resulting surface mixed layer is important for reasons ranging from the local phytoplankton dynamics to the climate system. Even though the mixed-layer depth (MLD) is a parameter of great interest,[2] there is little agreement on how it should be defined. This is partly because the most appropriate definition can depend on the scientific application. For example, studies of sound propagation might use a definition in terms of sound speed (Helber et al. 2008), while studies of mixing might motivate a definition in terms of turbulence (Brainerd and Gregg 1995). The characteristics of the data are also important, e.g. a method developed for a CTD that samples at high vertical resolution might fail when applied to the coarse output of numerical models.

This goal here is not to assess the various methods of inferring MLD from oceanographic data,[3] but rather to sketch roughly how R might be used in this sort of work.

One class of methods involves inferring the thickness of a near-surface region within which water properties are nearly constant. For example, the MLD may be defined as the shallowest depth at which density or temperature differs from the surface value by a fixed amount $\Delta\rho$ or ΔT. As noted in Table 1 of Kara et al. (2000), studies based on density commonly use $\Delta\rho = 0.125$ kg/m^3, while ΔT values as low as $0.1\,°C$ and as high as $1\,°C$ have been suggested, as well as a criterion on $\Delta T(\partial\rho/\partial T)$ being at most 0.5 kg/m^3. A temperature scheme could be set up as follows:

```
data(ctd, package="oce")
plotProfile(ctd, xtype="temperature", ylim=c(15, 0),
            col.temperature="black")
temperature <- ctd[["temperature"]]
pressure <- ctd[["pressure"]]
for (criterion in c(0.1, 0.5)) {
    inMLD <- abs(temperature[1]-temperature)
    < criterion
    MLDindex <- which.min(inMLD)
    MLDpressure <- pressure[MLDindex]
    abline(h=pressure[MLDindex], lwd=2, lty="dashed")
}
```

---

[2]The Sverdrup (1953) paper on the influence of mixed layer depth to phytoplankton biology has been cited nearly a thousand times to date.

[3]Suggestions for MLD inference abound in the literature. Useful recent treatments include those of Kara et al. (2000), Thomson and Fine (2003), de Boyer Montégut et al. (2004), Helber et al. (2008), Chu and Fan (2010a), and Chu and Fan (2010b). The last in this list discusses a method used in an exercise here.

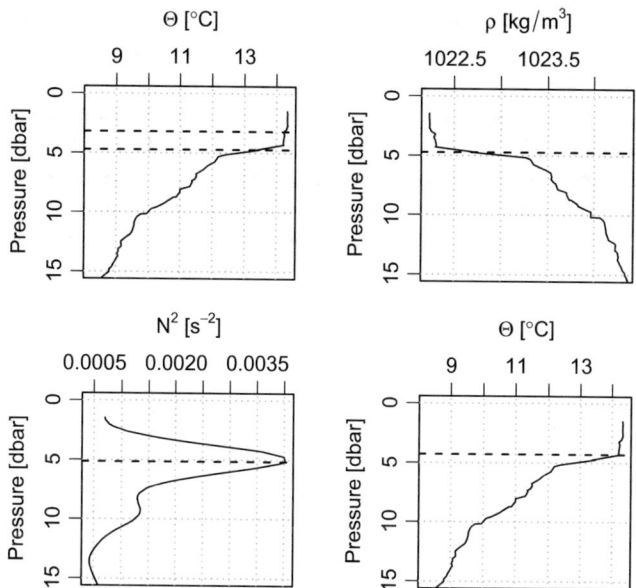

Fig. 5.4 Comparison of four methods for inferring mixed-layer depth. Top left: profile of Conservative Temperature, with mixed-layer depths inferred using temperature steps of 0.1 and 0.5 °C. Top right: similar, but for density step 0.125 kg/m³. Bottom left: based on the peak of N^2. Bottom right: based on the method of Chu and Fan (2010b)

The results (top-left of Fig. 5.4) are noticeably different for the two ΔT criteria. The deeper estimate is a good match with the results of the $\Delta\rho = 0.125 \, \text{kg/m}^3$ method (code not shown), as given in the top-right panel.

A second class of methods involves derivatives of water properties, based on, e.g., $\partial T/\partial z$ or $\partial^2 T/\partial z^2$. For example, the thermocline could be inferred as the spot where $|\partial T/\partial z|$ is largest, with the region above being interpreted as a mixed layer. A problem with such methods is that differentiating noisy signal produces results that can be difficult to interpret, so some smoothing may be required, which introduces another parameter that must be adjusted. One way to handle smoothing is with a smoothing spline, as is used by the oce function swN2() to estimate

$$N^2 = -\frac{g}{\rho}\frac{\partial\rho}{\partial z} \tag{5.1}$$

This could be used in a gradient-based method, e.g. the bottom-left panel of Fig. 5.4 is created with

```
plotProfile(ctd, xtype="N2", ylim=c(15, 0),
col.N2="black")
mid <- which.max(swN2(ctd))
pstar <- pressure[mid]
abline(h=pstar, lwd=2, lty="dashed")
```

The bottom-right panel of Fig. 5.4 shows the results of a method proposed by Chu and Fan (2010b), as implemented in the solution to Exercise 5.5. In it, the water column is examined at a sequence of depths below the surface. Denoting a given depth as z_*, say, the procedure is to perform a linear regression of temperature with respect to depth for all measurements lying above z_*, and then to use this regression to predict temperatures over some distance D below z_*. Chu and Fan (2010b) denote the standard deviation of the fit above z_* as E_1, and the absolute value of the mean temperature misfit within the test region below z_* as E_2. If z_* lies at the bottom of the mixed layer, E_1 will be small, while E_2 will be large. Thus the ratio E_2/E_1 may serve as an indicator of whether z_* is at the bottom of the mixed layer. Chu and Fan (2010b) suggest calculating the MLD as the depth z_* where E_2/E_1 achieves maximal value.

Exercise 5.5 Use the Chu and Fan (2010b) method on the `ctd` dataset. (See page 219 for a solution.)

5.2.2.7   Constructing Sections from CTD Profiles

There are several ways to archive hydrographic section data, and `oce` can handle the common ones.

WOCE archival data may be stored in a single CSV file, or in a directory containing several such files, one for each station. An example of the first case is the source of the `oce` dataset named `section`, which was constructed from a North Atlantic section denoted A03 in the WOCE system. This was created with `read.section()` with the `file` argument, as follows:

```
url <- "https://cchdo.ucsd.edu/data/7872/a03_hy1.csv"
section <- read.section(file=url, sectionId="a03",
        institute="SIO", ship="R/V Professor Multanovskiy",
        scientist="Vladimir Tereschenkov")
```

Reading a section stored as multiple CTD files is similar, with the `directory` argument naming a source directory. Another method is to read CTD objects individually (see Exercise 5.1), using `as.section()` to combine them.

5.2.2.8   Plotting Sections

The individual station profiles within oceanographic sections may not share a common set of sampling depths, but the `oce` package provides `sectionGrid()` to grid the data, so that contour or image plots can be created. The `section` section can be used to illustrate, as in Fig. 5.1 near the start of this chapter. The first step in creating this diagram is to load the data and subset by longitude using a value that retains the Gulf Stream region, with

```
data(section, package="oce")
GS <- subset(section, longitude < -70)
GS <- sectionSort(GS, by="longitude")
```
where the sorting operation will define distance on the contour plot in terms of the westernmost station. Although the `plot()` function will automatically grid these data, it may be preferable to specific a particular grid, e.g. using pressures 0, 25, 50, ..., 1600 dbar as follows:
```
GSG <- sectionGrid(GS, p=seq(0, 1600, 25))
plot(GSG, which=c(1, 99), map.xlim=c(-80, -60))
```
where the map view is adjusted to show a recognizable portion of the coastline. Further customization, e.g. drawing depth contours, is made easier by using separate `plot()` calls.

Exercise 5.6 Plot dynamic height and geostrophic velocity across the Gulf Stream. (See page 220 for a solution.)

Exercise 5.7 Plot a split-depth temperature section for the Gulf Stream, with a panel for variation in the top 200 m and another for variation below. Use `layout()` to make panels of unequal height. (See page 221 for a solution.)

5.2.3 Water-Mass Analysis

5.2.3.1 Inferring Water Types

Sverdrup et al. (1942) suggested using the phrase "water type" to refer to a homogeneous body of water. This appears as a single point on a temperature-salinity diagram, or any display of conservative water properties.[4] The signature of a "water mass" (see Sect. 5.2.3.2) is a curve on such a plot.

On a property–property plot, the point representing the passive mixture of two water types lies along a line connecting the types, and the position of the point on that line reflects the relative contributions of the two sources. This linear-mixing concept was applied by Östlund and Hut (1984), who used variations of $\delta^{18}O$ mass concentration to help infer water exchanges in the Arctic Ocean; this study provides a motivation for a simple illustration here (see, e.g., Tomczak 1999, for a broader framework).

Salinity will be denoted S, and $\delta^{18}O$ concentration X. Using $\mathbf{F} = (F_1, F_2, F_3)$ to indicate the mass fractions of the respective water types that make up a given product of mixing, the mass conservation equation is

$$F_1 + F_2 + F_3 = 1 \tag{5.2}$$

---

[4]Sverdrup et al. (1942) considered temperature a conservative property in this context because they suggested the omission of near-surface observations.

while that for salinity is

$$F_1 S_1 + F_2 S_2 + F_3 S_3 = S \tag{5.3}$$

and that for $\delta^{18}O$ is

$$F_1 X_1 + F_2 X_2 + F_3 X_3 = X \tag{5.4}$$

These are equations (1) through (3) of Östlund and Hut (1984), which may be rephrased in matrix form as

$$\mathbf{MF} = \mathbf{O} \tag{5.5}$$

where $\mathbf{M}$ is the known source property matrix defined by

$$\mathbf{M} = \begin{pmatrix} 1 & 1 & 1 \\ S_1 & S_2 & S_3 \\ X_1 & X_2 & X_3 \end{pmatrix} \tag{5.6}$$

$\mathbf{O}$ is the vector of observations

$$\mathbf{O} = \begin{pmatrix} 1 \\ S \\ X \end{pmatrix} \tag{5.7}$$

and $\mathbf{F}$ is the fraction vector

$$\mathbf{F} = \begin{pmatrix} F_1 \\ F_2 \\ F_3 \end{pmatrix} \tag{5.8}$$

The inference of $\mathbf{F}$ is the goal of the present analysis. If the source water types do not lie on a single line in S-X space, the determinant of $\mathbf{M}$ is non-zero and the solution is

$$\mathbf{F} = \mathbf{M}^{-1}\mathbf{O} \tag{5.9}$$

which may be expressed

```
F <- solve(M, O)                    # letter O, not zero
```
in R. To make this more concrete, the steps leading to Fig. 5.5 will be laid out with simulated data. The first step is to define water types. The values

```
Ss <- c(36, 31, 19)
Xs <- c(0.50, -1.7, -3.1)
```
are plotted as a polygon with

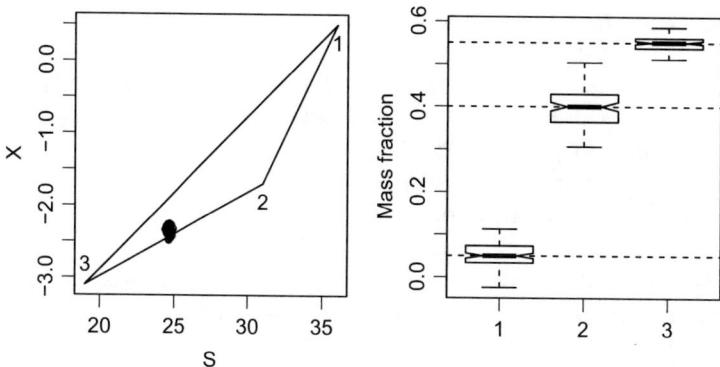

Fig. 5.5 Left: polygon connecting water types and dots for simulated data. Right: boxplots of inferred water fractions F_1, F_2 and F_3, with lines for unperturbed values (and negative values indicating points outside the polygon in the left panel)

```
plot(Ss, Xs, type="n", xlab="S", ylab="X")
polygon(Ss, Xs)
text(Ss, Xs, 1:3, pos=c(1,1,3))
```
in the left panel of Fig. 5.5, which also includes a "data" cloud drawn with
```
F <- c(0.05, 0.4, 0.55)
Sobs <- F[1] * Ss[1] + F[2] * Ss[2] + F[3] * Ss[3]
Xobs <- F[1] * Xs[1] + F[2] * Xs[2] + F[3] * Xs[3]
points(Sobs, Xobs, pch="+")
```
The matrix of source water properties, as in (5.6), may be constructed using rbind() to bind rows together.
```
M <- rbind(c(1, 1, 1), Ss, Xs)
```
A vector of observations is similarly created with
```
O <- rbind(1, Sobs, Xobs)          # letter O, not zero
```
after which solve() permits inference of the mass fraction vector **F′**
```
Fp <- t(solve(M, O))               # letter O, not zero
```
where t() puts the result in column form; accuracy is checked with
```
mean(abs(Fp - F))
| [1] 1.94289e-16
```
In this case, solve() gives tightly constrained results, because the polygon does not collapse onto a line.

The methodology can also be used to explore uncertainty, by perturbing the input data in Monte Carlo simulations. Those perturbations should reflect the measurement uncertainty, aspects of the sampling scheme, etc. For illustration, the values 0.5% for S and 2% for X will be used, with 100 perturbations.
```
n <- 100
err <- c(0.5, 2)
set.seed(5231)                     # for reproducibility
```

```
Fp <- matrix(NA, nrow=n, ncol=3)
for (i in 1:n) {
    Sp <- Sobs + rnorm(n=1, sd=abs(Sobs) * err[1]/100)
    Xp <- Xobs + rnorm(n=1, sd=abs(Xobs) * err[2]/100)
    points(Sp, Xp, cex=3/4)
    O <- rbind(1, Sp, Xp)
    Fp[i,] <- t(solve(M, O))
}
```

Various approaches may be used to investigate the Fp matrix thus calculated, and readers might want to consider histograms with hist(), density plots with density(), etc. A good balance between detail and overview is provided with boxplots, constructed with boxplot() as below, with the unperturbed values indicated with abline().

```
boxplot(Fp, notch=TRUE, ylab="Mass fraction")
abline(h=F, lty="dashed")
```

The fact that the horizontal lines on the right panel of Fig. 5.5 lie within the notches on the boxplots may be a reason for some confidence in this procedure, at least with the assumed simulation properties.

As the standard deviation was used to define uncertainties in S and X for the simulation, it may be used to infer the percentage uncertainties in the inferred source-water components, e.g.

```
round(100*sd(Fp[,1])/mean(Fp[,1]))
```

yields a 55% scatter in the inferred value of F_1, with similar calculations yielding 11% and 3% for F_2 and F_3, respectively. Another approach, suitable for actual data, might be to use a bootstrapping method (see Sect. 4.3).

5.2.3.2 Tracing Water Masses

The use of water-mass analysis to contextualize covariation of water properties in geographical or dynamical terms is a common theme of the literature (see, e.g., Tomczak 1999; Hinrichsen and Tomczak 1993; McDougall and Jackett 2007; Zika and McDougall 2008; Gebbie and Huybers 2010).

The section dataset can be used to illustrate how R can be applied to water-mass analysis. In this case, which involves careful study of property values, it is important to use handleFlags() to avoid the effects of spurious values (see Sect. 5.9.2), with

```
data(section, package="oce")
section <- handleFlags(section)
```

A good starting point may be to plot a temperature-salinity diagram, as in the top-left panel of Fig. 5.6, created with

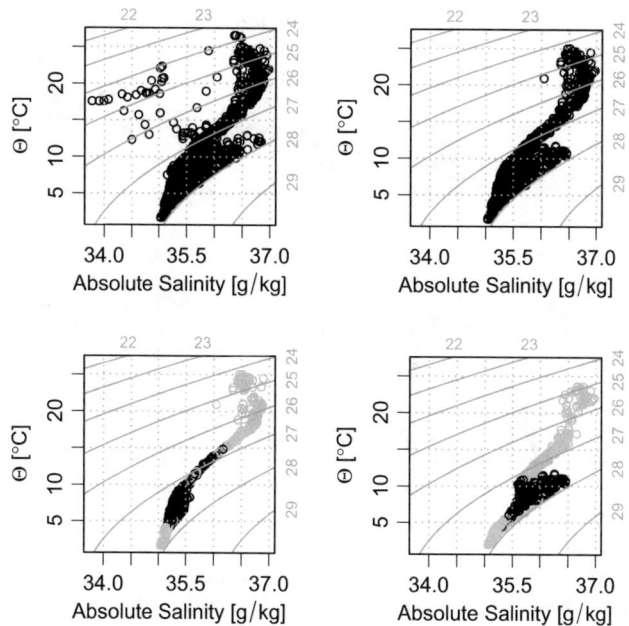

Fig. 5.6 T–S diagrams for the `section` dataset. Top: all observations (left) and just the open-ocean observations (right), showing the fresh anomalies to be in coastal waters. Bottom left and right: subdivision into observations west and east of 35W, with dark symbols for mid-depth waters

```
plotTS(section)
Slim <- par("usr")[1:2]
Tlim <- par("usr")[3:4]
```
where axis limits are stored for matching panels. Open-ocean stations are plotted in the top-right panel of Fig. 5.6 with
```
open <- subset(section, 9 < stationId & stationId<107)
plotTS(open, Slim=Slim, xaxs="i", Tlim=Tlim, yaxs="i")
```
where the most obvious feature is the absence of low-salinity coastal waters seen in the full dataset.

Readers familiar with the North Atlantic might wonder whether the warm and salty waters at density anomalies of 27.5–28 kg/m^3 are a signature of Mediterranean outflow. A simple way to check is to contrast waters to the west and east of the midpoint of the transect, with
```
W <- subset(open, longitude < -35)
E <- subset(open, longitude >= -35)
c1 <- "black"
c2 <- "gray"
plotTS(W, col=ifelse(abs(W[["pressure"]]-1000)<400,
```

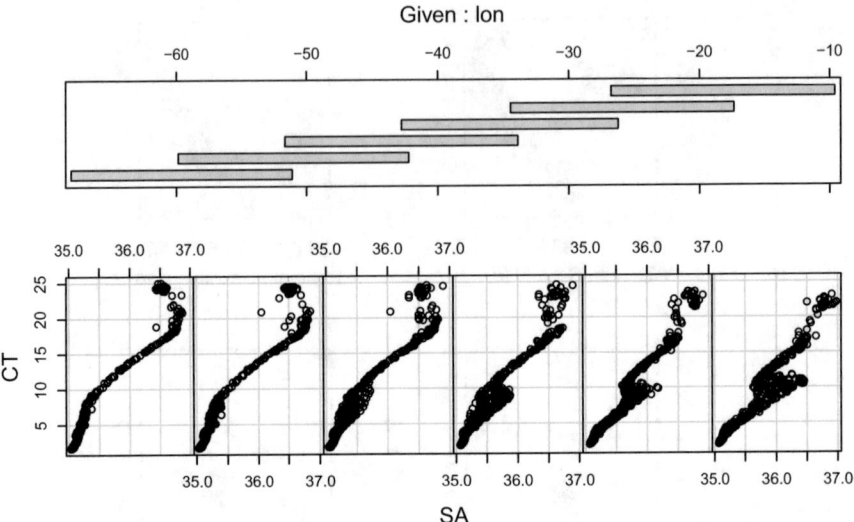

Fig. 5.7 Spatial variation of the temperature-salinity relationship for the open-ocean `section` data, showing the Mediterranean signature, increasing eastward

```
      c1, c2), Slim=Slim, xaxs="i", Tlim=Tlim, yaxs="i")
    plotTS(E, col=ifelse(abs(E[["pressure"]]-1000)<400,
      c1, c2), Slim=Slim, xaxs="i", Tlim=Tlim, yaxs="i")
```
where the colour specification highlights waters near 1 km depth. As expected, the resultant bottom panels of Fig. 5.6 show that the anomalously warm and salty waters are mainly in the eastern half of the domain. Further detail could be found by colour-coding for latitude, e.g.

```
    cm <- colormap(open[["longitude"]])
    drawPalette(colormap=cm)
    plotTS(open, col=cm$zcol, mar=c(3, 4, 2, 4))
```
(results not shown here) reveal that the Mediterranean signal is halved at about 30°W.

Multivariate relationships may also be shown with a conditioning plot, e.g.

```
    CT <- open[["CT"]]
    SA <- open[["SA"]]
    lon <- open[["longitude"]]
    coplot(CT~SA|lon, rows=1)
```
produces Fig. 5.7, the panels of which are temperature-salinity diagrams (isopycnals could be added via the `panel` argument) arranged by longitude, as indicated by the plots in the top margins. This gives a clear indication of the western decrease of the Mediterranean signature.

Exercise 5.8 Create a `coplot` of the `section` dataset, showing *T–S* dependence as a function of latitude and longitude. (See page 221 for a solution.)

Exercise 5.9 Suggest how to produce T–S diagrams categorized by longitude, using cut(), factor() and split. (See page 222 for a solution.)

5.2.3.3 Inferring Transport Mechanisms

The density ratio, which measures the relative contributions of temperature and salinity to the density stratification, is an important parameter for double diffusive convection (see, e.g., Turner 1973), and for the salt-finger mode of convection, the ratio is defined in the UNESCO equation of state by[5]

$$R_\rho = \left(\alpha \frac{\partial \theta}{\partial z}\right) \Big/ \left(\beta \frac{\partial S}{\partial z}\right) \tag{5.10}$$

where

$$\alpha = -\frac{1}{\rho_0} \frac{\partial \rho}{\partial \theta} \quad \text{and} \quad \beta = \frac{1}{\rho_0} \frac{\partial \rho}{\partial S} \tag{5.11}$$

are the thermal expansion coefficient and haline contraction coefficient, respectively. Ingham (1966) noted out that large regions of the world ocean have nearly uniform values of R_ρ, and Schmitt (1981) suggested that this might be a consequence of flux divergences associated with salt-finger convection. As part of his analysis, Schmitt rewrote (5.10) as

$$\theta' = \theta_0 + \int_{S_0}^{S} R_\rho(\beta/\alpha) \, dS \tag{5.12}$$

and applied a least-squares fit of θ' to the observed θ, with θ_0 and R_ρ taken as free parameters.[6] Reproducing his results in R is a simple matter of combining nonlinear regression with function integration.

The schmitt dataset in the ocedata package contains data for the North Atlantic example in Figure 1 of Schmitt (1981), and

```
data(schmitt, package="ocedata")
plotTS(as.ctd(schmitt$S, schmitt$theta, 0),
eos="unesco")
```

---

[5]The GSW equation of state expresses R_ρ analogously, but the UNESCO system is used in this section for easier comparison with the literature referenced here.

[6]The value of θ_0 depends on the value of S_0, but the latter may be chosen for convenience of plotting or matching the range of data.

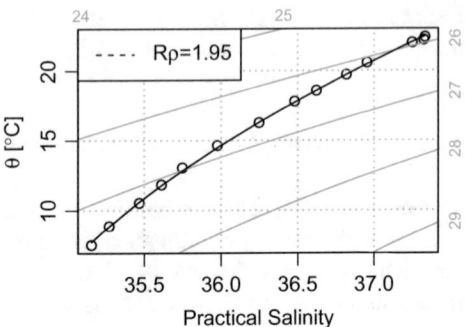

Fig. 5.8 Determination of R_ρ, emulating an analysis by Schmitt (1981)

creates a reproduction as Fig. 5.8 here. A function for $\partial\theta/\partial S$ is

```
dthetadS <- function(S, theta, parms)
    list(parms$Rrho * ( swBeta(S, theta, 0,
    eos="unesco") /
                        swAlpha(S, theta, 0,
                        eos="unesco")))
```

and so a model for θ is constructed by integrating this, with

```
library(deSolve)
thetaPrime <- function(theta0, Rrho)
    lsoda(theta0, schmitt$S, dthetadS,
    list(Rrho=Rrho))[,2]
```

where the first `lsoda()` argument is a starting value for the fit, the second gives values of S at which θ' is to be reported, the third is the function to be integrated, and the fourth is a list of parameters passed to that function. It now remains to fit the model with

```
thetaFit <- nls(theta~thetaPrime(theta0, Rrho),
                data=schmitt, start=list
                (theta0=5,Rrho=2))
lines(schmitt$S, predict(thetaFit))
```

where the first `nls()` argument specifies the model, the second is a data frame holding the data, and the third is a list of starting values. A legend is constructed with

```
Rrho <- coef(thetaFit)[2]
legend("topleft", lty="dashed", bg="white",
        legend=substitute(paste(R*rho,v),
                        list(v=sprintf
                        ("=%.2f", Rrho))))
```

Note that Schmitt (1981) inferred a value of $R_\rho = 1.948$, whereas the present analysis gives a 95% confidence interval $R_\rho = 1.95 \pm 0.03$.

Exercise 5.10 Formulate a model in which the misfit in S is minimized, and evaluate the confidence interval for it. (See page 223 for a solution.)

5.3 Acoustical Data

The oce package can handle Doppler data from both profilers and velocimeters. In most cases, the files may be read with read.oce(). For example,

```
d <- read.oce("../data/adp_rdi_3737.000",
from=3450, to=3550)
```

reads an RDI-Teledynedata file, with from and to being optional arguments specifying a range of profiles to be input. The resultant object has specialized versions of generic functions for subsetting, plotting, etc., and these make it easy to process the data.

As an example, plotting a time-series of pressure is a good way to find when the mooring reached the bottom, and

```
plot(d, which="pressure", type="p")
```

produces Fig. 5.9, which shows an increase from atmospheric pressure (near-zero, by convention for this instrument) to a sea pressure corresponding to about 10 m depth. A glance at this diagram indicates a settling time between 21:42 and 21:43, and this estimate could be narrowed further with, e.g.,

```
start <- numberAsPOSIXct(locator(1)$x);
abline(v=start)
```

(Note that the combination of actions with semicolon is convenient for interactive tasks of this type.)

Similar procedures can identify the end of the useful data, after which

```
dw <- read.oce("../data/adp_rdi_3737.000",
from=start, to=end)
```

could be used to read just the in-water data, or the entire file could be read and then subsetted, with, e.g.,

```
d <- read.oce("../data/adp_rdi_3737.000")
dw <- subset(d, start <= time & time <= end)
```

After trimming for time, the next step in the processing of Acoustic Doppler data often involves adjusting the velocity coordinates. Many groups record data in a non-Cartesian axis system that is aligned with the acoustical beams. Such data are said to be in "beam coordinates." The oce package supports this system, in addition to

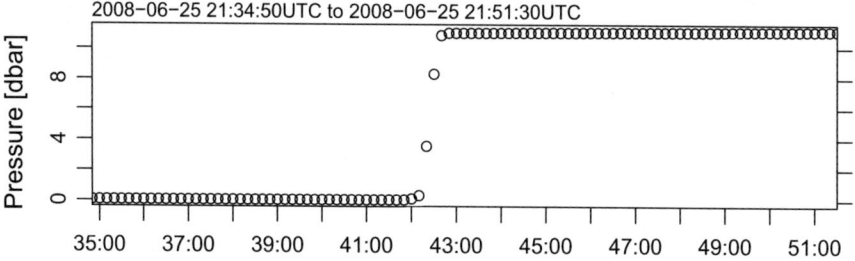

Fig. 5.9 A plot used in an interactive session to find a mooring-settling time

a Cartesian coordinate system called "xyz" that is referenced to the instrument, and another called "enu" that has components in the eastward, northward, and upward directions. Plots and calculations can be done in each of the coordinate systems, and functions are provided to convert between them.

For example, the oce package provides a sample ADCP dataset

```
data(adp, package="oce")
```

that was created by reading in beam coordinates with

```
adpBeam <- read.oce("adp_rdi_2615.000",
    from=as.POSIXct("2008-06-26", tz="UTC"),
    to=as.POSIXct("2008-06-27", tz="UTC"),
    by="10:00", latitude=47.88126, longitude=-69.73433)
```

and then converting to xyz coordinates and thereafter enu coordinates with[7]

```
adpXyz <- beamToXyz(adpBeam)
adp <- xyzToEnu(adpXyz)
```

The plot() function for ADCP objects has many variants, and these are selected with the which argument. If which is not supplied, the default is a stacked image plot, displaying the components of velocity and, for four-beam ADCPs, an "error" velocity related to the mismatch of vertical velocity computed in two different ways. Other values of which provide for dozens of plot varieties, including which="pressure", as used above.

A common step in most analyses is to plot the data. Of the many ways to do this, simple overviews are a common first step. For example,

```
data(adp, package="oce")
plot(adp, which="uv+ellipse")
```

produces the diagnostic plot of depth-integrated horizontal velocity components shown in the left panel of Fig. 5.10.

The author has high-resolution coastline and bathymetry data for this region, with which the right panel of Fig. 5.10 may be created with

```
load("../data/coastlineSLE.rda")
load("../data/sltopo.rda")
lat <- adp[["latitude"]]
lon <- adp[["longitude"]]
plot(coastlineSLE, clatitude=lat,
clongitude=lon, span=30)
lines(z030$longitude, z030$latitude, lty=3) # 30m
u <- apply(adp[['v']][,,1], 1, mean, na.rm=TRUE)
v <- apply(adp[['v']][,,2], 1, mean, na.rm=TRUE)
scale <- 1 / cos(lat * pi / 180)
points(lon+u*scale/30, lat+v/30, cex=1/2)
```

thus revealing a tendency of these tidal currents to be aligned with bathymetry.

---

[7]Actually, the oce function toEnu() calculates ENU values without intermediate steps, but for detailed work it is common to study each variant for clues about issues such as flow anomalies caused by the mooring, magnetometer problems, etc.

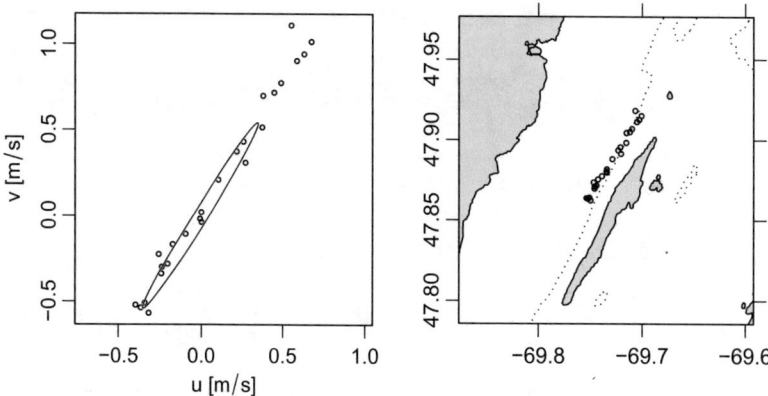

Fig. 5.10 ADCP plot showing currents with covariance ellipse (left) and currents in relation to coastline and 30 m isobath (right)

5.4 Sea-Level and Tidal Analysis

The `sealevel` dataset from the `oce` package holds observations made in Halifax Harbour during the year 2003, and it can be used to illustrate some common tasks in sea-level analysis. For example,

```
data(sealevel, package="oce")
plot(sealevel)
```

creates Fig. 5.11, in which the top panel shows the whole data set, the middle panel a month-long sequence, and the bottom panel a log-power spectrum with some tidal constituents marked.

The spike in September indicates a storm surge caused by Hurricane Juan, one of the most damaging recorded storms in this area (Xu and Perrie 2012). The first step in isolating a storm-surge signal is to remove the tidally forced variations in sea level, and this is handled with the `oce` function `tidem()`, which fits tidal models with methods[8] similar to those of the popular Matlab program called `T_TIDE`, described by Pawlowicz et al. (2002) and the earlier Fortran code described by Foreman (1977). A tidal model is inferred with

```
m <- tidem(sealevel)
```

---

[8]For more information on tidal analysis, see, e.g., Munk and Cartwright (1966), Godin (1972), Pugh (1987), and Foreman and Neufeld (1991).

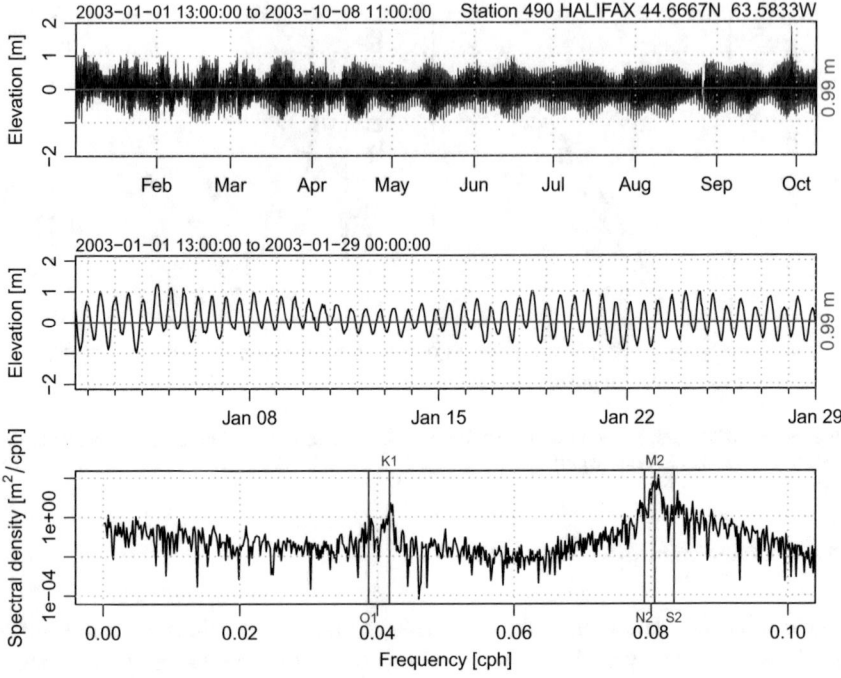

Fig. 5.11 Sea-level time series measured in 2003 in Halifax Harbour

which, in this default form, automatically selects tidal constituents based on the sampling interval. The non-tidal residual of sea level is found with

```
eta <- sealevel[["elevation"]] - predict(m)
```

and the view can be focussed on the event[9] as in Fig. 5.12 with

```
par(mfrow=c(2,1))
tlim <- as.POSIXct(c("2003-09-24","2003-10-05"),
tz="UTC")
plot(sealevel, which=1, xlim=tlim)
abline(v=as.POSIXct("2003-09-29 04:15:00", tz="UTC"),
lty=2)
t <- sealevel[["time"]]
oce.plot.ts(t, eta, xlim=tlim, ylab=expression(eta),
xaxs="i")
abline(v=as.POSIXct("2003-09-29 04:15:00",tz="UTC"),
lty=3)
```

---

[9]Hurricane Juan made landfall at Halifax about midnight local time (0400h UTC). The author recalls the moment when the deformations of his window glass had suddenly switched from being fascinating to terrifying.

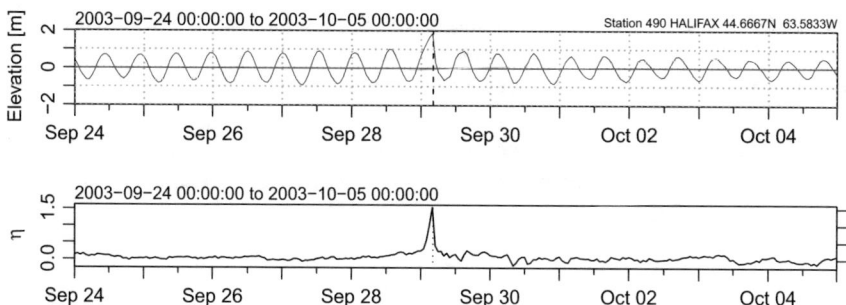

Fig. 5.12 Halifax sea level during Hurricane Juan. Top: observations. Bottom: deviation from a tidal model. The dotted line indicates the arrival of Juan at Halifax

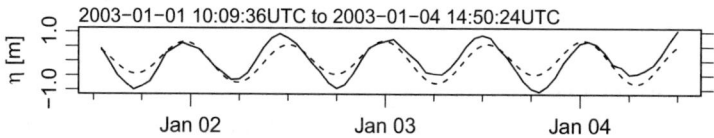

Fig. 5.13 Sea level in Halifax Harbour (solid) along with WebTide prediction (dashed)

the > 1 m storm surge was worsened by occurrence near high tide.

Many applications also benefit from the incorporation of dynamical models of tides. One example is WebTide, available from the Canadian Department of Fisheries and Oceans,[10] which provides a GUI scheme for retrieving data from tidal atlases of sea level and currents. The oce package provides an R interface to the WebTide database, e.g.

```
lat <- sealevel[["latitude"]]
lon <- sealevel[["longitude"]]
t <- sealevel[["time"]][1:72]
eta <- sealevel[["elevation"]][1:72]
eta <- eta - mean(eta)
oce.plot.ts(t, eta, ylab=expression(eta*" [m]"))
tr <- range(t)
tw <- seq(tr[1], tr[2], by="15 min")
e <- webtide("predict",lat=lat,lon=lon,time=tw,
plot=FALSE)
lines(e$time, e$elevation, lty="dashed")
```

produces Fig. 5.13, showing Halifax Harbour data, along with WebTide sea-level predictions.

Exercise 5.11 Compare the spring-neap variation in Halifax sea level with the phase of the moon. (See page 223 for a solution.)

---

[10]www.bio.gc.ca/science/research-recherche/ocean/webtide/index-eng.php.

Exercise 5.12 Use the `oce` function `plotTaylor()` to contrast three tidal models of sea level in Halifax Harbour: one with default constituents, one with just M2, and one with just S2. (See page 224 for a solution.)

Exercise 5.13 Determine whether the storm surge from Hurricane Juan in Halifax Harbour can be detected after removing tidal energy with a Doodson tidal filter. (See page 225 for a solution.)

5.5   Coastlines

Many formats of coastline data files are recognized by `read.oce()` or variants of `read.coastline()`. Cartesian plots are obtained with, e.g.,

```
data(coastlineWorld, package="oce")
plot(coastlineWorld)
```

and map projections (see Appendix C) are obtained either by providing the `projection` argument to `plot()` or to `mapPlot()`.

5.6   Topography

Topographic (bathymetric) data can be plotted as contours, e.g.

```
data(topoWorld, package="oce")
plot(topoWorld)
```

or as images, e.g.

```
imagep(topoWorld)
imagep(topoWorld, colormap=colormap(name="gmt_globe"))
```

where the latter employs a common topographic palette.

Projected images are handled with `mapImage()`, e.g.

```
mapPlot(coastlineWorld, projection="+proj=wintri")
mapImage(topoWorld, colormap=colormap
(name="gmt_globe"))
```

Exercise 5.14 Construct a hypsometric curve using `outer()` to create an area matrix that pairs with the depth matrix `topoWorld[["z"]]`. (See page 225 for a solution.)

5.7   Argo Floats

The `read.oce()` function detects the NetCDF format of Argo data, e.g.

```
argo <- read.oce("../data/6900388_prof.nc")
```

reads the data recorded by the float with ID number #6900388, designated as the January 2011 "Float of the month" by the British Oceanographic Data Centre. This dataset is provided in the oce package, accessed with

```
data(argo, package="oce")
```

and the overview provided with

```
summary(argo)
```

reveals that there are 210 delayed-mode profiles and 13 corrected profiles in this dataset. The former may be plotted with

```
argoD <- subset(argo, dataMode=="D")
plot(argoD)
points(argo[["longitude"]], argo[["latitude"]], pch=20)
```

the result of which will reveal some differences from Fig. 2.6 on Page 42.

Exercise 5.15 Graph surface water properties for the delayed-mode portion of the argo dataset, as a T–S diagram and a trajectory with symbol size proportional to spiciness. (See page 226 for a solution.)

5.8 Satellites

Specialized software is typically used to analyse satellite data, but R can help with many tasks, providing benefits relating to flexibility of processing and integration of different data types. The landsat package (Goslee 2011) supports Landsat 5 and Landsat 7 datasets, with functions for such things as data input, plotting, conversion from engineering units to physical units, image registration, and cloud detection. At the time of writing, the landsat package was unable to deal with the data from newly launched Landsat 8 satellite, so some basic support was added to the oce package, as will be illustrated with a temperature calculation here.

The oce package provides a sample Landsat 8 dataset named landsat. This is a wintertime view of Nova Scotia and surrounding waters, decimated to a 1 km grid to speed processing. All 11 wavelength bands are available, but the present focus is restricted to tirs1, the first band of the thermal infrared sensor. Spanning wavelengths from 10.6 to 11.2 μm, this band may be used to estimate temperature,[11] making it broadly relevant to oceanography.

Landsat data are stored in two-byte values (or "counts") that must be converted to physical units. The left panel of Fig. 5.14 shows the result of estimating the at-satellite brightness temperature using

---

[11] The tirs2 band could be used similarly, with different coefficients in the formulae. Also note that those coefficients are subject to change; see, e.g., Barsi et al. (2003) for insights with respect to previous satellites and the USGS site http://landsat.usgs.gov/calibration_notices.php for calibration notices for Landsat 8.

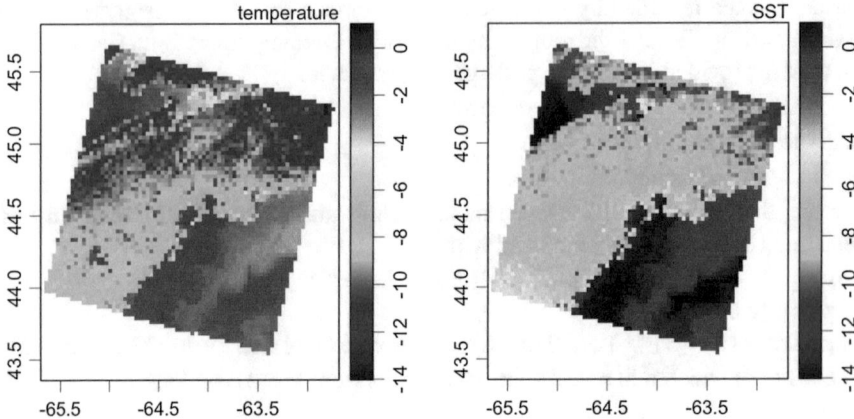

Fig. 5.14 Landsat-8 image of at-satellite brightness temperature (left) and an estimate of surface temperature based on a freezing-temperature offset (right)

$$T_B = \frac{K_2}{\ln(1 + K_1/\lambda_L)} - 273.15 \qquad (5.13)$$

where d is the `tirs1` count and $\lambda_L = A_L + M_L d$, and A_L, M_L, K_1 and K_2 are calibrated values stored in a header file (see the USGS website[12]). These things may be written as

```
data(landsat, package="oce")
d <- landsat[["tirs1"]]
AL <- landsat[["header"]]$radiance_add_band_10
ML <- landsat[["header"]]$radiance_mult_band_10
lambdaL <- AL + ML * d
K1 <- landsat[["header"]]$k1_constant_band_10
K2 <- landsat[["header"]]$k2_constant_band_10
TB <- K2 / log(1+K1/lambdaL) - 273.15
TB[d == 0] <- NA
```

but this is somewhat tedious, so `oce` provides an accessor function to carry out the calculation, using a pseudo-band named `"temperature"`, and

```
plot(landsat, band="temperature",
     col=oce.colorsJet, zlim=c(-14, 1))
```

yields the left panel of Fig. 5.14 a cloud-free day in March, 2014. The land is much colder than the ocean, and readers familiar with the topography of Nova Scotia

---

[12]http://landsat.usgs.gov/Landsat8_Using_Product.php.

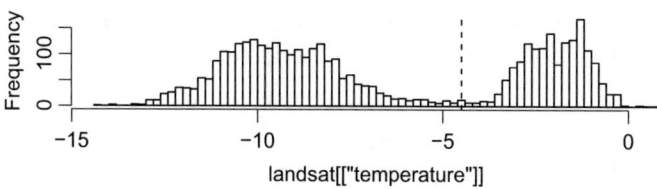

Fig. 5.15 Brightness temperature, with a line indicating a possible water-ice cutoff

might also note how temperature drops with altitude, and how frozen lakes appear as homogeneous patches.[13]

Patterns in the open ocean are consistent with knowledge of the region, e.g. with ~ 10 km swirls south of Nova Scotia suggesting baroclinic eddies. However, the patterns are different in the Bay of Fundy, where a granular texture, colder temperatures, and local knowledge all suggest the presence of sea ice. The nearby waters have brightness temperature T_B ranging from -4 to $-3\,°C$, and the fact that this is well below the freezing temperature demonstrates that further adjustment is required to infer surface temperature.

The calculation of surface temperature requires knowledge of local surface emissivity and atmospheric properties (see, e.g., Jiménez-Muñoz et al. 2014). This is beyond the present scope, but procedures may be illustrated with a more speculative approach. Assuming waters near sea ice to be near the freezing temperature, a temperature offset could be computed with a study of subregions that show water. This would require close inspection of the images, but the methodology can also be illustrated from a histogram, e.g.

```
hist(landsat[["temperature"]], breaks=100, main="")
TBice <- -4.5
abline(v=TBice, lty="dashed")
```

yields Fig. 5.15, in which the vertical line shows a reasonable dividing line between ice and water, suggesting that surface temperature is

```
SST <- TB - TBice - 1.8
```

where 1.8°C is the freezing temperature for typical oceanic salinity. This result may be stored within the landsat object with

```
landsat <- landsatAdd(landsat, SST, "SST")
```

and then plotted with

```
plot(landsat, band="SST", col=oce.colorsJet,
zlim=c(-14,1))
```

the results of which, sketched in the right panel of Fig. 5.14 but more apparent in a full-resolution image, have ocean temperature exceeding the freezing point.

---

[13]The statements about lakes and granularity of ice in the Bay of Fundy refer to the full-resolution Landsat data, not to the heavily decimated version in the landsat object.

Exercise 5.16 Use functions in the mixtools package to separate the histogram of Fig. 5.15 into two parts. (See page 227 for a solution.)

5.9 General Analysis

5.9.1 Processing QC Flags

It is conventional in oceanographic analysis to use quality-control (QC) flags to indicate questionable data values, instead of setting those values to NA. This has the benefit of nuance, but it can complicate analysis.

QC flag values may follow conventions set by manufacturers, by data-archiving agencies, or by individual analysts. Flags are usually associated with individual data entries, e.g. with a flag for each temperature in a CTD profile, along with separate sets of flags for the salinities and other properties.

Often, flags are integers whose value indicates the quality of the associated datum. A fundamental difficulty in dealing with flags is that different coding values are used for different instruments, and by different data-archiving agencies. For example, in the case of CTD data, the World Hydrographic Program (WHP) convention[14] is to use a flag equal to 2 for "acceptable" measurements, whereas the Canadian Department of Fisheries and Oceans (DFO) uses a flag equal to 2 to indicate "uncertain" data, in its website[15] providing data from the Atlantic Zone Monitoring Program. The argo convention is somewhat of an intermediate, with 2 indicating values deemed "probably good." Clearly, analysts must be careful when dealing with QC flags.

The oce package provides a simple scheme for handling QC flags, in a generic function called handleFlags() that has specialized variants for different data types. In usually replaces suspicious data with NA values (or alternative measurements stored in the data file), although analysts may supply functions to deal with problematic data in more sophisticated ways. Assumptions about flag conventions are made by handleFlags() based on the data class, but analysts may still need to tailor things for an individual dataset, as in the WHP versus DFO example stated above. For example,

```
data(section, package="oce")
stn100 <- section[["station", "100"]]
stn100[["phosphateFlag"]][14:16]
[1]  2 3 2
stn100[["phosphate"]][14:16]
[1]  1.02 1.15 1.10
```

---

[14]https://www.nodc.noaa.gov/woce/woce_v3/wocedata_1/whp/about.htm.

[15]http://www.meds-sdmm.dfo-mpo.gc.ca/isdm-gdsi/azmp-pmza/hydro/index-eng.html.

```
handleFlags(stn100)[["phosphate"]][14:16]
|[1] 1.02   NA 1.10
```

shows a case in which the default WHP flags will work, so the default action of handleFlags() is suitable. However, if these data had been in the DFO format, the appropriate call would have been

```
handleFlags(stn100, flags=list(2:4))
```

because DFO indicates problematic data with QC flags of 2–4.

Since summary provides statistics of QC flags, it is sometimes possible to *guess* the flags, on the assumption that good data are more common than bad data, but of course this is no substitute for studying the documentation that agencies provide with datasets.

For more on how CTD data are handled, type ?"handleFlags,ctd-method" in an R console. Similar queries provide help on other object types. The first part of the result is similar across oce classes, but tailored information is provided near the bottom of the output.

5.9.2 Handling Faulty Data

Although QC flags are often provided with archived oceanographic data, analysts at the forefront of research are often called upon to work with data that have yet to be examined for flaws, or for which standardized flags are judged to be questionable.

In examining data for problems, it can help to be aware of the challenges in making measurements at sea. Faulty data can be caused by measurement errors (clocks drift, sensors go out of calibration, biofouling causes instrument malfunction, etc.), instrumental setup errors (poor scaling factors can yield velocities that "wrap around" because of modulo arithmetic applied to small memory chunks, etc.) and platform limitations (tides can shift moorings, currents can knock down mooring lines, mooring-generated turbulence can occur for certain current directions, etc.).

Many problems are specific to the type of instrument, e.g. the salinity spiking problem of CTDs has no direct analogue in ADCP data, so it is difficult to provide general advice. Therefore, only two specific instruments are dealt in the next two subsections.

5.9.2.1 Diagnosing Data Faults Statistically

Although a thorough discussion of the problems with CTD data is beyond the present scope (see UNESCO 1988, for an introduction), a few examples may suggest how R can be helpful for this sort of work.

Returning to the discussion of Sect. 5.2.2.4, the ctdRaw dataset provides a convenient starting point, with the summary() revealing unphysical values for temperature and salinity, even though no flags have been set. The working

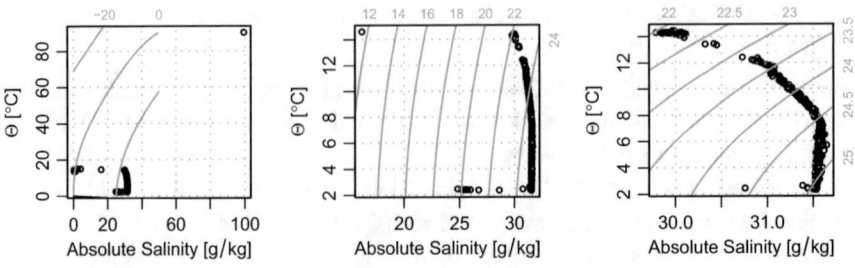

Fig. 5.16 Temperature-salinity relationship of the `ctdRaw` data in raw form (left) and after removing outliers in two passes (middle and right). The procedure removes most outliers, but retains a few suspicious points at low temperature

hypothesis of the previous treatment is that the odd values reveal a form of quality-control, even though an analysis of the header in the source file reveals the line

```
# bad_flag = -9.990e-29
```

which suggests otherwise. Such mixed messages are common, and dealing with them is a core part of practical operating procedures.

Oceanographers use flags of 99, 999, etc., because they are recognizable and outside the expected range of observation. This second aspect means that procedures to flag unphysical values will also detect such flags, and so it is worth explaining some methods used to detect unphysical values.

For example, the Seasoft CTD software can perform a "wild edit" procedure to identify suspicious values by departure from the mean by more than a specified multiple of the standard deviation (Sea-Bird Electronics 2016). This may be accomplished in R with

```
bad <- function(x, n=2)
    is.na(x) | (abs(x-mean(x,na.rm=TRUE))
    >n*sd(x,na.rm=TRUE))
```

and applying this to the data

```
o <- bad(ctdRaw[["salinity"]]) |
bad(ctdRaw[["temperature"]])
ctdRaw[["salinity"]][o] <- NA
ctdRaw[["temperature"]][o] <- NA
```

yields the T–S diagrams of Fig. 5.16. Note that using two passes, as recommended in the Seasoft manual, yields fewer spurious points than a single pass. Even so, a few suspicious points still stand out from the data cloud, as salinity reductions in the deep water. Readers experienced in hydrographic data might next turn to a study of anomalies in spiciness-density space. Another approach might be to look for rapid variations of properties with respect to pressure or sampling time. Detailed procedures might vary with the type (and state of calibration/repair) of the instrument, the wave state, etc. R makes it easy to do explore such approaches, while also freeing analysts from possible limitations of commercial software.

Exercise 5.17 Suggest a procedure for performing CTD outlier rejection using data flags, instead of modifying the data. (See page 227 for a solution.)

5.9.2.2 Diagnosing Data Faults Based on Scientific Context

Sometimes, statistical outlier detection is not the best approach. An example is provided by a snippet from a ten-day SontekADV recording made during the St Lawrence Estuary Internal Wave Experiment (Richards et al. 2013).

```
load("../data/adv-bad.rda")
```

This instrument was attached firmly to a weighted mooring sitting on the ocean bottom, so the compass heading should be nearly constant.[16] However, summary(d) indicates a range of nearly 200 degrees (in addition to suspicious values of roll). Since extrema are sensitive to outliers, it makes sense to look also at the summary statistics

```
quantile(adv[["heading"]])
    0%    25%    50%    75%   100%
   0.0    0.0   90.0  161.7  199.8
```

and these results are also inconsistent with expectation. In addition, Fig. 5.17 reveals that the heading shifts dramatically in fractions of a second, which is not physically possible for such a setup. Another sign of a problem is that the heading record is not continuous, but mainly shifts between particular values. A simple inference might settle on an angle of $0°$, since this value occurs so often, but a cautious analyst should be on the lookout for zero values, which may result from problems in connections between sensors and loggers. Such considerations underline the fact that blindfolded statistical analysis is no substitute for contextualizing data in physical terms.

R makes it very easy to explore data with such things in mind. For example, a crude function to remove data in a specified range, and calculate a median() (or other function, here named fcn) on the remaining data, is

```
fix <- function(x, bad, fcn=function(x) median(x))
    fcn(subset(x, x < bad[1] | x > bad[2]))
```

and this is used to exclude near-zero points with

```
h <- fix(adv[["heading"]], c(-2, 2)) # ignore near zero
par(mfrow=c(1,2))
hist(adv[["heading"]], main="", breaks=100)
abline(v=h, lty="dashed")
label <- sprintf("%.0f", h)
```

---

[16] An exception to this expectation of nearly constant headings occurs when the compass is aligned at the "cut point" of $0°$ or $360°$. The angleRemap() function from the oce package can be helpful if calculations are to be done on such angles.

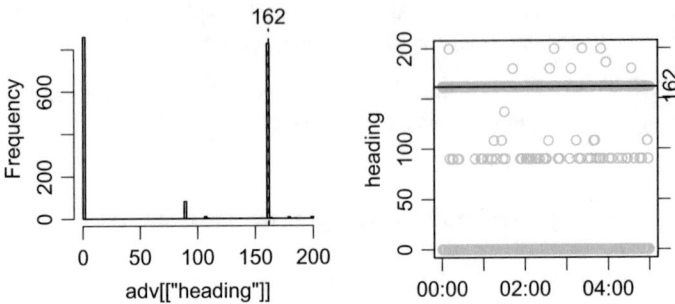

Fig. 5.17 Five minutes of heading data from an ADV that had an intermittently faulty orientation unit. Although the actual heading of the instrument is known to be lie between 100° and 200°, it would be difficult to infer this value from the histogram or the time series, given the frequent occurrence of zero values. The value in the margins is the heading calculated by a combined physical/statistical method explained in the text

```
mtext(label, side=3, at=h)
plot(adv, which=16, type="p", col="gray",
     drawTimeRange=FALSE)
mtext(label, side=4, at=h)
abline(h=h, lwd=2)
```

resulting in Fig. 5.17. An analyst who considered h to be preferable to the existing heading record might decide to update the object, with

```
adv <- oceSetData(adv, "heading",
                  rep(h, length((adv[["time"]])))),
                  note="replace zero-contaminated
                  heading")
```

5.9.3   Dealing with Log-Normally Distributed Data

Log-normally distributed quantities are of interest in many fields (Johnson et al. 1995; Limpert et al. 2001). An important oceanographic example is the rate of viscous dissipation of turbulent kinetic energy, ϵ, which is used (Osborn 1980) to infer K_V (Sect. 4.4). One way to distill a set of ϵ measurements is to calculate $E_1 = \exp(\mu)$, where μ is the mean of $\ln \epsilon$. This yields an estimate of the population median. A corresponding estimate of the mean is given by $E_2 = \exp(\mu + \sigma^2/2)$ with σ being the standard deviation of $\ln \epsilon$ (see, e.g., Johnson et al. 1995, Chapter 14). In the context of averaged dynamical equations, E_2 is of more direct utility than E_1. The contrast between E_1 and E_2 can be illustrated with artificially constructed data, e.g.

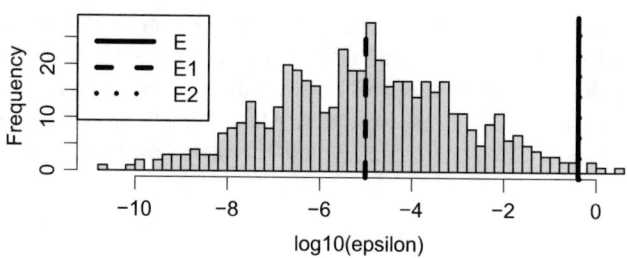

Fig. 5.18 Characterizing dissipation rate ϵ with median (E1) and mean (E2) measures; E shows the mean of the source distribution used to generate the random values shown

```
lmuHat <- -5 * log(10)            # converted to ln()
lsigmaHat <- 2 * log(10)          # converted to ln()
set.seed(593)                     # for reproducibility
epsilon <- rlnorm(500, lmuHat, lsigmaHat)
hist(log10(epsilon), main="", col="lightgray",
breaks=50)
```

creates the histogram in Fig. 5.18. The second and third arguments to `rlnorm()` are natural logs, so a factor is applied to represent $\widehat{\mu} = 10^{-5}$ and $\widehat{\sigma} = 10^2$. The theoretical mean $E = \exp(\widehat{\mu} + \widehat{\sigma}^2/2)$ and the quantities E_1 and E_2 are also indicated in Fig. 5.18 with

```
E <- exp(lmuHat + lsigmaHat^2/2)
E1fcn <- function(x) exp(mean(log(x)))
E2fcn <- function(x) exp(mean(log(x)) + 0.5
* sd(log(x))^2)
abline(v=log10(E), lwd=5)
abline(v=log10(E1fcn(epsilon)), lwd=5, lty=2)
abline(v=log10(E2fcn(epsilon)), lwd=5, lty=3)
legend("topleft", lwd=5, lty=1:3, seg.len=3,
        legend=c("E", "E1", "E2"))
```

In this trial, E_2 is much to E than E_1 is, suggesting that the `E2fcn()` offers the better solution for tasks requiring the mean value.

Exercise 5.18 Compare E_1 and E_2 in a series of random trials. (See page 228 for a solution.)

5.9.4 Time-Series Analysis

Time series analysis is used widely in many fields, including economics, engineering, and both the social and natural sciences. The topic is deep enough to demand the scale of a textbook, popular examples of which include classics by Jenkins and Watts (1969) and Box and Jenkins (1976), and newer treatments by Brillinger (1981), Priestley (1981), Bloomfield (2005) and Shumway and Stoffer (2006). The last of these deals with time series in R, touching on wavelet analysis, state-space modelling, and a host of advanced topics. All of the above are aimed at general audiences; see, e.g., Chapter 5 of Emery and Thomson (2001) for an oceanographic context.

5.9.4.1   Time-Series Objects in R

The `ts()` function creates a time-series object, of class `"ts"`, given data sampled at a *constant rate*.

Given that the interval between samples is constant, complete information about the sampling times can be stored within the object using just three attributes. For example, a sequence of 5 Weibull-distributed random data might be created with

```
ts(rweibull(5, 1))
Time Series:
Start = 1
End = 5
Frequency = 1
[1]  0.4450218 0.6482199 5.3843208 1.2436122 0.8863650
```

Here, the start time, end time, and frequency have all taken on default values, but using extra arguments to `ts()` provides great flexibility in specifying sampling times, e.g.

```
ts(rweibull(5,1), frequency=0.1)
# start    1, end 41
ts(rweibull(5,1), start=0, frequency=0.1)
# start    0, end 40
ts(rweibull(5,1), end=10, frequency=0.1)
# start -30, end 10
```

produce results summarized in the comments, but

```
ts(rweibull(5, 1), start=1, end=10, frequency=0.1)
```

produces an error, because of the contradiction in the arguments for a time-series of length 5.

The main advantage of time-series objects is that they contain information about the sampling times, which is generally useful (e.g. in plotting) but also specifically useful for important methods such as the computation of spectra (introduced in Sect. 2.4.10, with more discussion in Sect. 5.9.4.5, below). However, since many oceanographic data are not sampled at a uniform rate, it makes sense to begin with a discussion of methods for dealing with irregular data.

5.9.4.2 Interpolation Methods for Nearly Regular Data

The simplest cases are those in which the sampling rate is *nearly* constant, e.g. for logging instruments that usually record properly but that occasionally produce bad data owing to electronic problems or sensor limitations. For such problems, a reasonable approach may be to use an interpolating function to re-sample based on times. A simple approach is to use `approx()` for linear interpolation between points, e.g. for artificial data with two bad intervals

```
t <- 1:20
s <- sin(2*pi*t/10)
s[c(5:7, 15)] <- NA
```

a new version of the signal could be constructed with

```
ok <- is.finite(s)
tok <- t[ok]
sok <- s[ok]
sNew <- approx(tok, sok, t)$y
```

Exercise 5.19 Use `approx` to write a function that interpolates from one set of times to another. (See page 228 for a solution.)

5.9.4.3 Windowing Methods for Irregularly Sampled Data

Creating regularly spaced data by using gap-filling procedures can be problematic for data collected at widely ranging sampling intervals. For example, interpolating across a narrow gap to replace a spike (see Sect. 5.9.4.6) is unlikely to alter a spectrum greatly, but doing so across a missing season of hourly observations can cause deceptive spectral reddening.

A common alternative to gap-filling is to forgo high-frequency resolution, creating a new time series by averaging within fixed time intervals. For example, measurements made every few days might be averaged within 1-month windows. Such procedures yield a fixed "sampling" interval, which then opens the door to conventional time-series analysis. They also improve statistical reliability of samples, which can be desirable even with regularly sampled data, if low-frequency variability is the primary interest (recall the discussion of bin-averaging CTD data in Sect. 5.2.2.5).

There is no single agreed-upon method for this work. The windows may be distinct, or they may overlap. The data within windows may be treated equally, or there may be a focus near window centres. The number of data in each window may be recorded, perhaps for use in the weighting of models to be fitted later, or it may be discarded. The computation done within a window may be based on value alone or on a model of variation over time, e.g. using `lm()` predictions at central times can avoid problems of data unevenly distributed within windows of trending data.

Fig. 5.19 GISS time-series (gray) with pentadal average (thick black) and confidence interval (thin black)

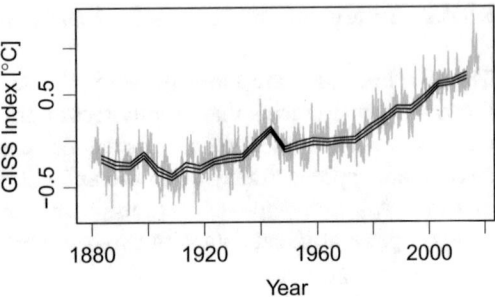

In any case, the first step is a subdivision of the data in time categories, and this is accomplished easily with the `cut()` and `split()` functions explained in Sect. 2.3.7, after which, e.g., the base function `apply()` or `laply()` from the `plyr` package can be used for the computation. For example, Fig. 5.19 shows the application of a running-mean scheme to the Goddard Institute for Space Studies (GISS) land-ocean temperature index, which is stored as the `giss` dataset of the `ocedata` package.[17] The first step in constructing this figure is to plot the raw data:

```
data(giss, package="ocedata")
y <- giss$year
i <- giss$index
plot(y, i, type="l", col="gray",
     xlab="Year",ylab=expression
     ("GISS Index ["*degree*"C]"))
```

after which `cut()` may be used to set up, say, a pentadal averaging scheme (with `ceiling()` and `floor()` being used for yearly window boundaries)

```
C <- cut(y, breaks=seq(ceiling(min(y)),
   floor(max(y)), 5))
```

which is then used to split the data into 5-year chunks

```
ys <- split(y, C)
is <- split(i, C)
```

within which the data are averaged

```
library(plyr)
ymean <- laply(ys, mean)
imean <- laply(is, mean)
```

and plotted as the thick line in Fig. 5.19 with

```
lines(ymean, imean, lwd=2)
```

while the next exercise shows how to draw the thin lines, indicating a confidence interval.

---

[17]The `giss` data are on a uniform sampling interval, but the procedure given here would work identically if this were not the case.

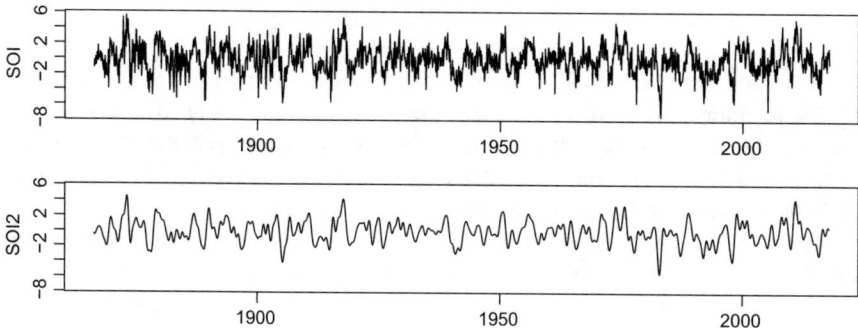

Fig. 5.20 Southern Oscillation Index time series since the year 1880 in original form (top) and after smoothing with a low-pass Butterworth filter (bottom)

Exercise 5.20 Add confidence limits to Fig. 5.19. (See page 229 for a solution.)

5.9.4.4   Time-Domain Analysis

The Southern Oscillation Index (SOI), provided as `soi` in the `ocedata` package, is a well-known example of a time series that has variability on a range of time-scales, but that lacks a prominent overall trend.[18] This can be seen in the top panel of Fig. 5.20, made with

```
data(soi, package="ocedata")
SOI <- ts(soi$index, start=soi$year[1], frequency=12)
plot(SOI, xlab="")
```

using the generic function for plotting a `ts` objects. The signal has a ragged appearance that somewhat obscures the interannual variations that are of great interest in the context of weather patterns. One way to highlight those variations is to low-pass filter the data, removing higher-frequency variations.

There are two basic approaches to filtering. Denoting the signal of interest x_i, where i is an index to time, the filter output might be defined using a non-recursive filter (also called a finite impulse response filter) in which the output y_i depends only on the input[19]

$$y_i = a_0 x_i + a_1 x_{i-1} + \ldots \tag{5.14}$$

or a recursive filter (also called an infinite impulse response filter), in which previous values of the output are used as well as input values

---

[18] The relevance of trends for filtering is discussed at the end of this section.

[19] Terms for, e.g., x_{i+1} can be included in the term indicated with ellipsis, yielding non-causal filters.

$$y_i = a_0 x_i + a_1 x_{i-1} + \cdots + b_1 y_{i-1} + b_2 y_{i-2} + \ldots \tag{5.15}$$

A review by Harris (1978) provides a detailed comparison of several non-recursive filters that were in common use in the 1970s. Many of these are still employed routinely today, although recursive filters are also popular. Amongst the latter, the filter of choice for many oceanographers is the Butterworth filter (see, e.g., Roberts and Roberts 1978), so it is worth illustrating here.

The `signal` package, patterned on a popular Matlab system, supports the creation and use of a wide variety of filters. Its `butter()` function creates coefficients for Butterworth filters of low-pass, high-pass, band-stop, and band-pass varieties. The filter critical frequencies are specified in its `W` argument, as multiples of the Nyquist frequency, e.g. to design a filter with cutoff period τ_c, the appropriate value of `W` is $2\tau_s/\tau_c$, where τ_s is the time between samples. For example, the following creates a filter with a 12-month cutoff that may be applied to the monthly sampled `soi` data:

```
library(signal)
filt <- butter(n=4, W=2*1/12)
# W=2*tauS/tauC in months
```

The return value from `butter()` is a list containing vectors a and b that hold the coefficients defined in (5.15). To apply the filter, it is common practice to use `filtfilt()`, which runs the filter both forward and backward in time in order to cancel out the phase distortions that occur with recursive filters. It is important to note that `filtfilt()` squares the spectral transfer function, so that the 3 dB reduction at the cutoff frequency changes to a 6 dB reduction (see Exercise 5.23). Applying the filter to the `soi` time series with

```
SOI2<-ts(filtfilt(filt, soi$index), start=1866,
frequency=12)
plot(SOI2, xlab="", ylim=par("usr")[3:4], yaxs="i")
```

produces the lower panel of Fig. 5.20, where the `ylim` and `yaxs` settings copy the y-axis geometry from the previous plot. There is good evidence of interannual variability in this low-passed view.

It should be noted that recursive filters such as the Butterworth variant must make an assumption about the filter output, i.e. the y values in (5.15), "before" the start of the time series, which can yield spurious predictions at the start of the output. If `filtfilt()` is used to do the filtering, the problem can also occur at the end of the time series. The effect is most pronounced when the input values are far from zero, as for data with significant trends. Exercise 5.22 deals with a crude method for addressing this endpoint problem. Other schemes have also been proposed; see Gustafsson (1996) for an entry into the literature on such methods for the Butterworth case and, e.g., Mann (2008) for a more general discussion of endpoint problems with other filters, framed in a climate-change context.

Exercise 5.21 Following Sect. 2.4.8, use `acf()` to look for oscillations in the `soi` dataset. (See page 229 for a solution.)

Exercise 5.22 With the `giss` dataset, show how detrending can reduce spurious endpoint effects of Butterworth filters. (See page 229 for a solution.)

5.9.4.5 Spectral and Wavelet Analysis

Time-domain filtering can be useful for isolating variations within frequency bands that are known to be pertinent to particular applications, but it is not an ideal tool for exploring variation across a range of frequencies. Tasks in the latter category can be better handled with tools such as autocorrelation analysis and spectral analysis. These have been outlined in Sects. 2.4.8 and 2.4.10, and the goal of the present section is to expand on the latter, continuing with the `SOI` time-series constructed in Sect. 5.9.4.4.

A spectrum plot of `soi` may be created with a simple `spectrum()` call

```
spectrum(SOI, log="no", main="")
abline(v=1/c(2, 8), col="gray")
```

as shown in the top-left panel of Fig. 5.21, with gray lines indicating periods of 2 and 8 years. The results suggest a broad band of energy at periods of several years, which is reminiscent of El Niño Southern Oscillation timescales (see, e.g., Philander and Fedorov 2003), although the raggedness makes it difficult to assess levels. The

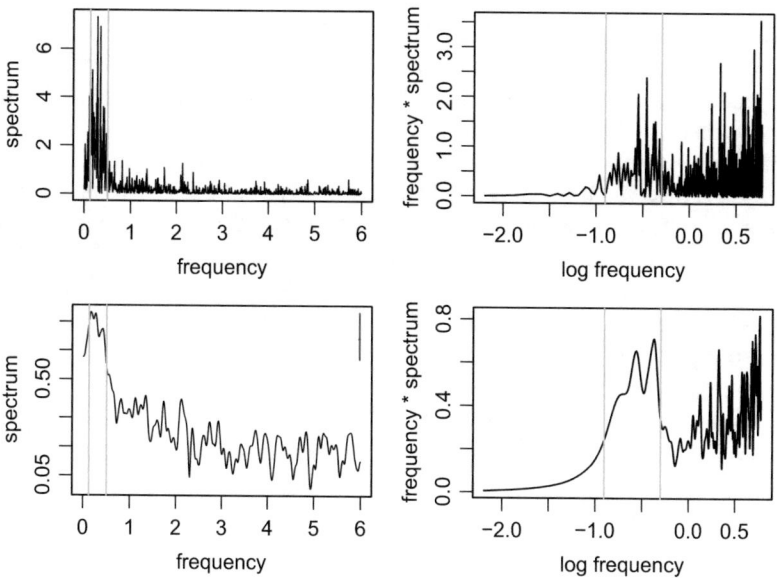

Fig. 5.21 Spectra of `soi` dataset, with frequency in cycles/year. The left panels show raw and smoothed spectra in conventional form, while the right panels show a variance-conserving form with logarithmic frequency. Gray lines mark periods of 2 and 8 years

pattern is clarified somewhat by smoothing in frequency space, and this may be controlled with the spans argument, e.g.

```
spectrum(SOI, spans=c(11, 9, 5), main="")
abline(v=1/c(2, 8), col="gray")
```

yields the lower-left panel of Fig. 5.21, showing a clearer view of energy distribution through the year-to-decade band. Variation in this time range can be investigated more easily if frequency is log transformed. If the vertical axis is altered to show the product of power and frequency, the result is a variance-conserving spectrum, in which the ratio of the area under two plotted peaks equals the ratio of variance in the two frequency bands (see, e.g., Glover et al. 2011, Section 6.3). Such representations of the raw and smoothed SOI spectra are shown in the right panels of Fig. 5.21, constructed with

```
s <- spectrum(SOI, plot=FALSE)
plot(log10(s$freq), s$freq*s$spec, type="l",
     xlab="log frequency", ylab="frequency
     * spectrum")
abline(v=log10(1/c(2, 8)), col="gray")
```

for the raw spectrum, and

```
ss <- spectrum(SOI, spans=c(11, 9, 5), plot=FALSE)
plot(log10(ss$freq), ss$freq*ss$spec, type="l",
     xlab="log frequency", ylab="frequency
     * spectrum")
abline(v=log10(1/c(2, 8)), col="gray")
```

for the smoothed spectrum. Each of the panels reveals significant variation in the 2 to 8 year band, with detail (such as the hint of two peaks) being easier to discern in the smoothed log-frequency case.

This comparison highlights the usefulness of smoothing spectra. The goal is usually to have enough smoothing to give acceptable confidence in the spectral level, but not so much that relevant peaks are smoothed away (for a much more expansive treatment, see Jenkins and Watts 1969, Section 7.2). Since a basic understanding of filtering is assumed here, the problem reduces to learning what spans does, and kernel() helps greatly with that. For example, providing spectrum() with spans=2 yields a smoothing kernel

```
kernel("modified.daniell", 2)
```

which has coefficient values $(1/8, 1/4, 1/4, 1/4, 1/8)$, i.e. a boxcar filter with endpoints that have been halved, which is known as a modified Daniell filter. If spans is a vector containing more than one element, the resultant smoothing filter is created by convolving the filters that are constructed for each value in turn, e.g. convolving filters of length 3, 5 and 9

```
plot(kernel("modified.daniell", c(3, 5, 9)))
```

yields the left panel of Fig. 5.22; comparing panels reveals how convolving sharp-edged filters yields a smoother result, and that careful selection of sub-filter lengths can control the width of the flat pass band; see, e.g., Shumway and Stoffer (2006) for more on such issues.

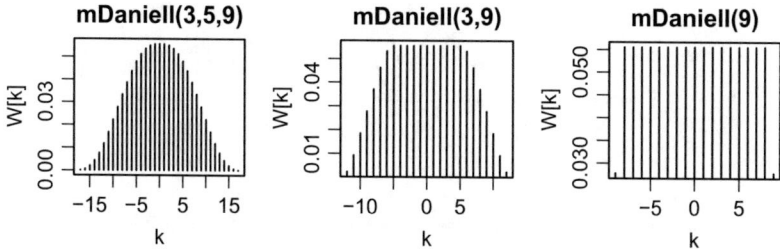

Fig. 5.22 Examples of spectral filters, corresponding to supplying `spectrum()` with `spans=9`, `spans=c(3,9)`, or `spans=c(3,5,9)`, from left to right

Fig. 5.23 Frequency-time plot, analogous to a sonogram or spectrogram, for constructed chirp signal plus noise (Exercise 5.24)

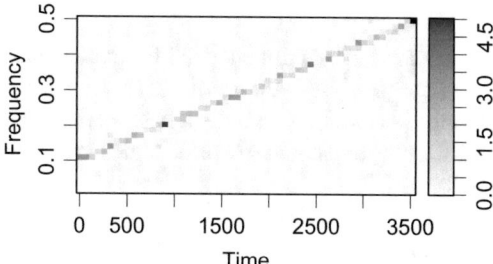

This R approach to spectral smoothing may be unfamiliar to some readers, who are accustomed to using the popular `pwelch()` function in Matlab, which divides the time series into sub-intervals and averages the spectra of the components (Welch 1967). This scheme is also provided by the `oce` package, with a more limited function of the same name.

It should be noted, however, that subdivision of time series is useful for more than averaging. It also opens the door to studies of how the spectral character of a system varies over time. This can be important in a wide range of applications, an example in the author's research being the phasing of internal wave incidence with respect to the tide, which may reveal forcing mechanisms and propagation pathways.

Using `cut()` and `split()`, it is a simple matter to subdivide a time series into segments for individual spectral analysis. Exercise 5.24 addresses this, leading to the construction of Fig. 5.23 for artificial data. Diagrams of this general sort are sometimes called spectrograms, and they have been used in oceanography since the early days of time-series measurement, e.g. for sonograms of whale calls (Schevill and Watkins 1962) and for the Snodgrass et al. (1966) study of swell propagation across ocean basins.

A modern expansion on the traditional frequency-time plot is the wavelet plot. A key feature in this analysis is the use of non-repeating basis functions for the spectral decomposition. A commonly used basis function is the so-called Morlet function illustrated in Fig. 5.24 and defined by

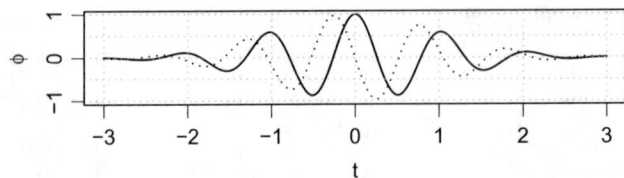

Fig. 5.24 Real (solid) and imaginary (dotted) components of a Morlet wavelet, $\phi = \phi(t, f)$ for nondimensional frequency $f = 6$

$$\phi = \left(e^{-ift} - e^{-\frac{1}{2}f^2}\right)e^{-\frac{1}{2}t^2} \tag{5.16}$$

and where f is nondimensional frequency and t is translated and nondimensionalized time (see, e.g., Ashmead 2012). Roughly speaking, wavelet analysis applies a sliding time window to a dataset in an examination of similarity to basis functions of different frequencies (Morlet et al. 1982; Daubechies 1990; Torrence and Compo 1998).

The `WaveletComp` package (Roesch and Schmidbauer 2014) handles wavelet calculations and their graphical display with simple functions,[20] e.g. the temporal variation of spectral energy in the SOI dataset is calculated using

```
library("WaveletComp")
data(soi, package="ocedata")
date <- seq(ISOdatetime(soi$year[1],1,1,0,
0,0,tz="UTC"),
            length.out=length(soi$year),by="1 month")
w <- analyze.wavelet(data.frame(date=date,
index=soi$index),
                "index", dt=1/12, lowerPeriod=1)
```

where the `dt` value indicates that frequencies are to be reported in cycles per year, and the `lowerPeriod` value removes sub-yearly oscillations from the analysis. The resultant value is an object of class `"analyze.wavelet"`, and its components may be further analysed and plotted in several ways. A useful overview plot such as Fig. 5.25 may be created with

```
wt.image(w, plot.ridge=FALSE, siglvl=0.005,
            color.key="i", show.date=TRUE)
```

where the second argument prevents the drawing of ridge lines, which can be distracting on a complex diagram, the third sets a significance level, and the fourth indicates that power is to be shown, as opposed to quantile. The diagram indicates that the main SOI variability is in the interannual band, and also that the energy in that band varies on decadal timescales, consistent with visual inspection of Fig. 5.20.

---

[20]The `biwavelet` package is an alternative.

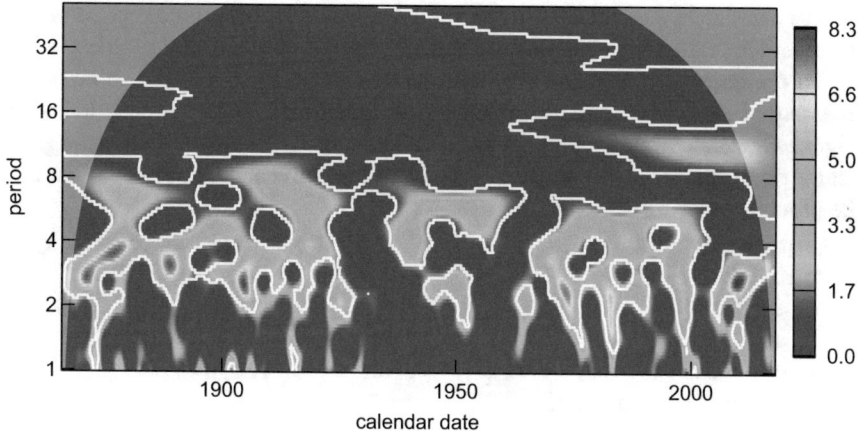

Fig. 5.25 Wavelet spectrum of the `soi` dataset. The light-coloured regions in the top corners indicate the difficulty of inferring long periods near the ends of a time series. Contour lines surround regions satisfying a significance criterion, and colour indicates spectral power

Exercise 5.23 Use `spectrum()` to demonstrate the Butterworth filter response. (See page 230 for a solution.)

Exercise 5.24 Write a function to compute and display a frequency-time plot. (See page 231 for a solution.)

Exercise 5.25 Use `fft()` to compute rotary spectra as defined by Gonella (1972). (See page 232 for a solution.)

5.9.4.6 Despiking Time-Series Data

A common problem in dealing with oceanographic time-series data is the handling of "spikes," i.e. short-lived and high-amplitude departures from otherwise more smoothly varying signals. Even though special considerations apply to different data types (see, e.g., Weekley et al. 2010), some general approaches are worth illustrating.

Methods for spike detection typically combine statistical analysis and an understanding of the characteristics of the instrument, the setting, and the quantity being measured. For example, in hydrostatic flow, the time derivative of velocity is small compared with the acceleration due to gravity, so a spike criterion in a velocity series $u = u(t)$ might be written

```
bad <- (abs(diff(u) / diff(t))) > (criterion * 9.8)
```

with `criterion` being adjusted to suit the application.

It can be helpful to combine variables in a dynamically meaningful way, rather than to consider all measurements individually, and also to work with primary variables instead of derived variables (e.g. focussing on temperature and conductivity, rather than temperature and salinity).

Purely time-series approach can also prove useful. An example is to develop a measure of the departure of the observations from a smoothed representation of the signal. Especially if this is done in combination with a measure of event brevity, this can yield results that are similar to visual spike identification. An illustration may be provided with artificial data, e.g.

```
set.seed(5946)                          # for reproducibility
par(mfrow=c(1, 3))
n <- 200
t <- 1:n
x <- exp(-t/100) + exp(-t/150) * sin(t/10)
+ rnorm(n, sd=0.05)
spiked <- c(50, 100, 101, 150)
x[spiked] <- x[spiked] + c(0.5, 0.5, 0.5, -0.5)
```

creates a signal that varies on two timescales and has both a random component and 4 added spikes.

Plotting this signal as Fig. 5.26 with

```
plot(t, x, type="l")
points(t[spiked], x[spiked])
```

shows that the points are not anomalous with respect to overall x values, since they lie within the range of the rest of the data. This means that a test of overall deviation will not be helpful in this case, as it was in the examination of the ctdRaw dataset. However, the points stand out locally from a smoothed version of x, as can be seen by examining the histogram of departure from such a smoothed version, e.g.

```
xs <- lowpass(runmed(x, k=11), n=5)
hist(x-xs, breaks=100, main="")
```

yields the middle panel of Fig. 5.26, showing outliers at high and low values. In some instances, an analyst might set cutoff criteria based on visual inspection of a histogram, but it can also be helpful to take the number of data into consideration, based on the probability of finding outliers of a given departure from the mean, e.g.

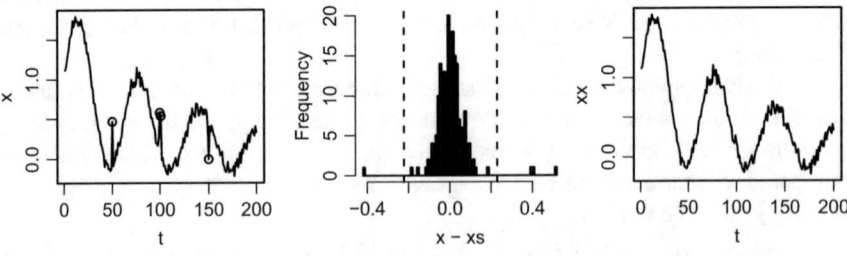

Fig. 5.26 Despiking a signal using statistics of departure from a smoothed curve. Left: signal $x = x(t)$ with spikes (circled) at $t = 50$, 100, 101, and 150. Middle: histogram of x minus x_s, a smoothed version of x. Right: result of replacing anomalous points with x_s

```
A <- 1
dev <- x - xs
lambda <- A * qt(1 - 1/(2*n), df=n-1)
abline(v=mean(dev) + sd(dev) * lambda*c(-1, 1), lty=2)
```

where A is a parameter that can be adjusted to increase or decrease the rejection rate. Now, a logical index of departure can be defined as

```
bad <- which((abs(dev - mean(dev))
> (lambda * sd(dev))))
```

For this constructed time series, the resultant `bad` is the same as `spiked`. This suggests some merit to this method, and explains its similarity to `despike()` in the `oce` package. However, the lifetime of the inferred spikes has not been considered explicitly, and that might be a useful variable to explore, in any further development along these lines.

Once spikes have been identified, there are three choices for further action: replace the points with `NA`, set flags to indicate concern, or replace the suspicious values with constructed values. In the case of replacement, a simple scheme is to fill the gap by interpolating linearly between adjacent values. This prevents extrema within the gaps, which is desirable in some cases (e.g. for density profiles, where there are physical reasons to distrust inversions on large length scales). Another common approach is to use a smoothed signal, which is already available for the scheme just described, i.e.

```
xx <- x
xx[bad] <- xs[bad]
plot(t, xx, type="l")
```

as is shown in the right panel of Fig. 5.26. The following example introduces more sophisticated statistical approaches.

Exercise 5.26 Use `na.kalman` from the `imputeTS` package to replace the spikes in x with the predictions of a Kalman filter. (See page 233 for a solution.)

5.9.4.7 Time-Series Forecasting

Oceanographers practice two types of forecasting with time series data. The first involves phenomena that are understood well enough that the prediction can have a dynamical component. Examples include the prediction of tides based on elevation records, using harmonic constituents of known frequency (Sect. 5.4) and the use of numerical ocean models to predict pollutant dispersal (see, e.g., Bourgault et al. 2014). In such cases, there is obvious benefit in tailoring the forecasting scheme to the dynamical situation at hand.

By contrast, the second type involves systems in which there is insufficient dynamical understanding (or pertinent data) to develop practical forecasts based on dynamical principles. General forecasting tools, which treat data in isolation from theory, may be helpful in such cases. A common method is simple regression, e.g.

extending a linear trend, or some other functional form, into the future. R is ideal for this sort of work, as discussed in Sects. 2.5.5 and 2.5.5.2. It also provides functions for fitting more complex stochastic models to data, and these merit an overview here.

A powerful stochastic method was described in the influential textbook by Box and Jenkins (1976), so it is often called the Box-Jenkins method. It involves the idea of a black-box system driven by white noise and described by a finite number of parameter values. For example, an autoregressive (AR) process is described by[21]

$$X_t = c + \sum_{i=1}^{p} \theta_i X_{t-i} + \epsilon_t \tag{5.17}$$

where t refers to the time step, c is a constant, and the θ terms are parameters of the process, as is the standard deviation of ϵ, which is assumed to be a white-noise time series with zero mean.

An autoregressive model with p terms in the sum is denoted $AR(p)$. A similar notation, $MA(q)$, is used for a moving-average model of the form

$$X_t = \mu + \sum_{i=1}^{q} \theta_i \epsilon_{t-i} + \epsilon_t \tag{5.18}$$

Such models can be combined into so-called ARMA models. If first-difference terms are added, an ARIMA (auto-regressive integrative moving average) model results.

A common forecasting analysis starts with inferring model parameters from an examination of a time series, after which the model is stepped forward in time to predict future variation. R provides several functions for this work. Here, the focus is on the arima() function, which fits ARIMA models.

The method will be illustrated with the adp dataset provided by oce, even though this is actually a case in which a tidal model could be used. The first step is to extract a component of velocity

```
data(adp, package="oce")
whichDepth <- which.min(abs(adp[["distance"]] - 20))
eastward <- adp[["v"]][,whichDepth,1]
```

and then plot it as a time series

```
ndata <- length(eastward)
npred <- 100
plot(eastward, type="l", lwd=2, xlim=c(0,
ndata+npred))
```

---

[21]Different treatments use different notations, and even disagree on signs; see ?arima for the R convention on the latter.

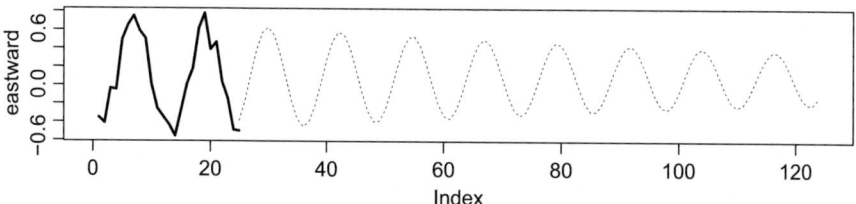

Fig. 5.27 Demonstration of ARIMA modelling

As the resultant Fig. 5.27 shows, velocity displays a tidal character, with added high-frequency variability.

A model with three auto-regressive coefficients and two moving-average coefficients may be constructed with

```
m <- arima(eastward, c(3,0,2))
```

where the middle value of the second argument specifies no differencing term, i.e. the function is being used in ARMA mode.

The print() methods produces a useful summary:

```
print(m)
```

```
Call:
arima(x = eastward, order = c(3, 0, 2))

Coefficients:
          ar1      ar2     ar3     ma1     ma2   intercept
        1.248   -0.142   -0.48   -1.241   0.241     0.076
s.e.    0.246    0.422    0.24    0.518   0.290     0.008

sigma^2 estimated as 0.0138:log likelihood=13.82,
aic=-13.65
```

and the forecast drawn in Fig. 5.27 with

```
index <- ndata + seq(0, length.out=npred)
p <- predict(m, n.ahead=npred)$pred
lines(index, p, lwd=0.5, lty="dashed")
```

reveals that the model captures the oscillating character of the data, while decaying towards zero as the data fall further into the past. Insight can be gained by repeating the procedure with different order values, and with constructed datasets of varying character; see Box and Jenkins (1976, Chapter 6) for more specific advice.

5.9.5 Gridding and Spatial Mapping

5.9.5.1 Challenges in Oceanographic Gridding

Whether acquired with ships, floats, gliders or moorings, oceanographic data are seldom arranged on a simple grid. Since gridded fields are useful for plotting, for gradient computation, and for use in finite-difference numerical models, gridding is a common task in oceanographic analysis. The work falls into the category of the analysis of spatial data, a topic addressed in several different R packages; see a "task view" on the R website.[22]

Although aspects of the work are general, it is prudent to keep the scientific context in mind, to avoid unrealistic results (see, e.g., Walters 1969). For example, when gridding hydrographic data for initializing a numerical model, it may make sense to select a method that yields results that are reasonably consistent with model dynamics, with density being gravitationally stable, etc. The end use of gridded fields should also be kept in mind, e.g. a contour diagram may need more smoothing than an image, because the eye is better at ignoring image speckle than a lacework of contour lines.

There are also considerations relating to the sampling patterns. Consider CTD profiles made along a ship track, for which sampling is dense in the vertical coordinate (sub-metre for raw data) and very much sparser along the ship track (often tens of kilometres).[23] These different scales call for different approaches. In the vertical, it is common to bin-average and then decimate to fixed depths (Sect. 5.2.2.5), often with attention being paid to density inversions. Profiles may then be combined in an along-track coordinate to yield gridded data, as is done by sectionGrid() in the oce package.

The combination of separate ship (or glider) tracks raises a variety of concerns owing to undersampling in space and time together. The ocean is so unsteady that naive gridding between ship tracks occupied years or decades apart can yield spurious results. Seasonality can be a particular problem, especially near the surface, but aggregating data reasonably can help with this, e.g. with atlases based on monthly climatologies (Levitus 1982; Levitus and Boyer 1994).

Analysts would be wise to become familiar with any gridding methods that are particular to their own fields of study, in order to take advantage of useful customizations of general methods (or to become aware of errors made by others). For example, Schmidtko et al. (2013) provide valuable insights on the gridding of hydrographic data in isopycnal and mixed-layer space. Smoothing is an important part of that analysis, as it is of most water-column studies. By contrast, some geological gridding methods use tessellated interpolation to respect each datum fully, as a way to avoid such problems as smoothing over fault lines (Sambridge et al. 1995).

---

[22]http://cran.r-project.org/web/views/Spatial.html.

[23]The resolution contrast holds for Argo floats as well as for ships, but not for ocean gliders.

In addition to such topical reading, analysts should study the broader literature, perhaps starting with Chapter 14 of Venables and Ripley (1999) and Chapter 4 of Emery and Thomson (2001). Other good starting points include Bretherton et al. (1976), Carter and Robinson (1987), and Davis (1985).

It should be noted that the statistical nature of the general gridding literature is somewhat divorced from the oceanographic reality. For example, mid-ocean ridges are effective barriers to water exchange, and so currents steered geostrophically along opposite sides of a ridge can have markedly different hydrographic properties. A person contouring a deep hydrographic field would take this into account, in calculating local averages "by eye." Ways to improve algorithms are addressed by Dunn and Ridgway (2002).

5.9.5.2   Least-Squares Gridding Methods

The `MASS` and `spatial` packages provide data and functions that are useful for illustrating least-squares spatial gridding, and analysts should strive to understand the basic approaches these packages provide.

For example,

```
library(MASS)
library(spatial)
data(topo)                          # in MASS package
```

yields `topo`, a data frame containing elements x, y, and z. The `surf.gls()` function in the `spatial` package may be used to fit surfaces with a generalized least-squares method, e.g.

```
glsModel <- surf.gls(np=2, covmod=expcov, x=topo, d=1)
```

where np is the degree of the model being fitted, `covmod` is a function used to calculate covariance (see the documentation for `expcov()` in the `spatial` package), x holds the (x, y, z) data, and d is a range parameter used by `expcov()`.

Now, it is a simple matter to set parameters, perform the calculation,

```
glow <- 0
ghigh <- 6.5
gn <- 50
glsPred <- prmat(glsModel, glow,ghigh,glow,ghigh,gn)
```

and contour the results (Fig. 5.28).

```
levels <- seq(0, 1000, 50)
contour(glsPred, levels=levels, labcex=1)
```

(The dashed contours in this figure are developed with a method explained in the next section.)

Fig. 5.28 Gridding the topo dataset, with the solid and dashed contours inferred from surf.gls() and interpBarnes(), respectively. The isolated numbers are percentage errors for the interpBarnes() method (see Exercise 5.27)

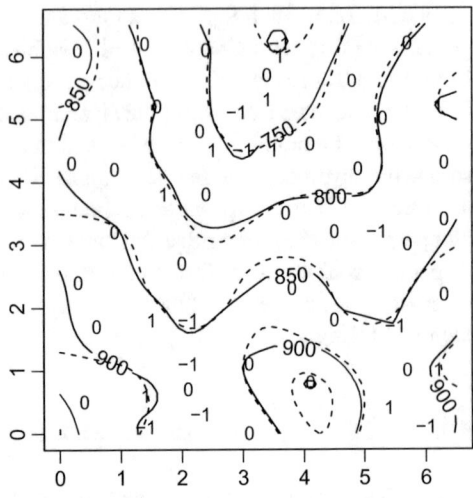

5.9.5.3 Weighted-Average Gridding Methods

The surf.gls() results of the previous section can be compared easily with those provided by the iterative method described by Barnes (1994), variants of which are used in NOAA oceanographic atlases (see, e.g., Levitus 1982; Levitus and Boyer 1994).

```
b <- interpBarnes(topo$x, topo$y, topo$z,
                  xg=seq(glow, ghigh, length.out=gn),
                  yg=seq(glow, ghigh, length.out=gn))
contour(b$xg, b$yg, b$zg, lty="dashed",
        levels=levels, drawlabels=FALSE, add=TRUE)
```

As shown in Fig. 5.28, the contoured fields from the two methods are similar, especially in regions of relatively high data density. Since there is no theoretical basis for either method, it is difficult to argue for one method over the other, but there is value in using several methods on problems like this, partly to get a measure of what might be called methodological uncertainty.

Exercise 5.27 Use the zd element of the return value from interpBarnes() to add the percentage errors to Fig. 5.28. (See page 233 for a solution.)

5.9.5.4 Bin-Averaging Before Gridding

Gridding large datasets can be computationally expensive. The time required to cast n data points onto a grid with N_x cells in one dimension and N_y in the second becomes proportional to $n N_x N_y$ for large values of n. With grid spacing chosen to match the data, N_x and N_y are each proportional to $\sqrt{n}$, and the time estimate becomes n^2 (see Appendix E for more on estimating computational cost). Such a rapid increase of cost with increasing dataset size can cause problems in moving from small test cases to practical applications.

A common solution to this problem, used, for example, in NOAA atlases, is to reduce the size of the data set by averaging and decimating within spatial bins. This can be done with a single pass through the data, i.e. with computation time proportional to n. The procedure can yield such dramatic reductions in computation time that it is the first step in many analyses of large oceanographic datasets.

Another compelling reason to bin average before gridding is to reduce biases resulting from uneven sampling density. For example, if waters within a harbour are sampled much more frequently than waters in the nearby ocean, the gridded ocean results may be biased towards the harbour value. Averaging in spatial bins could distill the harbour measurements to a single value, yielding more faithful representations of the nearby ocean. A similar approach can be used in the time domain, sometimes in terms of climatologies, in which data within identical seasons or months are binned together.

Bin averaging in R may be accomplished by combining `cut()`, `split()` and `lapply()`; an example for a one-dimensional grid is

```
set.seed(5954)                    # for reproducibility
x <- runif(100)
f <- x / (1 + x)
unlist(lapply(split(f, cut(x, pretty(x))), mean))
     (0,0.2]   (0.2,0.4]  (0.4,0.6]  (0.6,0.8]    (0.8,1]
 0.07984517 0.22484299 0.32744359 0.40592662 0.47635258
```

although it can be more straightforward to use `binMean1D()` from the oce package (Exercise 5.11)

```
binMean1D(x, f)$result
 [1] 0.07984517 0.22484299 0.32744359 0.40592662
 0.47635258
```

For the two-dimensional case, `binMean2D()` is useful, as may be illustrated in gridding the `secchi` dataset in the oce package. The data set holds in excess of forty thousand observations, which is not especially large in oceanographic terms, but it is still large enough to make gridding with `interpBarnes()` be slower than desired, for interactive work.

A good first step is to show the data locations, with, e.g.,

```
data(coastlineWorldFine, package="ocedata")
mapPlot(coastlineWorldFine, grid=5, col="gray",
        longitudelim=c(-5, 20),latitudelim=c(50, 66),
        projection="+proj=lcc +lat_1=50 +lat_2=65")
data(secchi, package="ocedata")
mapPoints(secchi$longitude, secchi$latitude,
pch=20, cex=0.3)
```

producing the left panel of Fig. 5.29, which suggests that sample-density bias may be an issue, the points being much more densely packed in the Baltic Sea and eastern North Sea than in more oceanic regions.

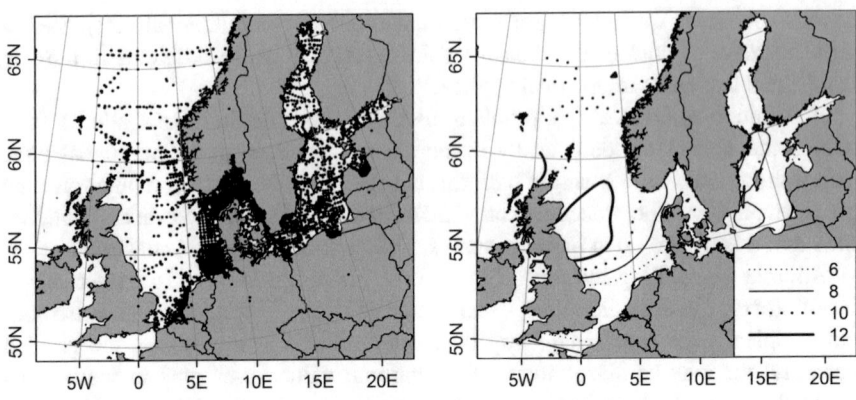

Fig. 5.29 Secchi depths. Left: data locations, drawn with Lambert Conformal Conic projection. Right: contours of bin-averaged and then gridded Secchi depth

A grid with simple longitude and latitude values may be constructed with

```
lonGrid <- pretty(secchi$longitude, 80)
# yields 1deg spacing
latGrid <- pretty(secchi$latitude, 40)
#      "   0.5deg   "
```

and then the bin averaging may be handled with

```
g <- binMean2D(secchi$longitude, secchi$latitude,
               secchi$depth, lonGrid, latGrid)
```

Before doing the gridding, it is necessary to convert the results into vectors, which can be done with the oce function ungrid()

```
u <- ungrid(g$xmids, g$ymids, g$result)
```

after which a smoothing scale is set (here, 3 times the decimation grid) and gridding (here, 100×100 geometry) is done with

```
asp <- 1 / cos(mean(u$y)*pi/180)
R <- 3 * diff(latGrid[1:2])
G <- interpBarnes(u$x, u$y, u$grid,
                  xr=asp * R * asp, yr=R,
                  xg=pretty(u$x, 100),
                  yg=pretty(u$y, 100))
```

The results are plotted as the right panel of Fig. 5.29 with

```
mapPlot(coastlineWorldFine, grid=5, col="gray",
        longitudelim=c(-5, 20), latitudelim=c(50, 66),
        projection="+proj=lcc +lat_1=50 +lat_2=65")
levels <- c(6, 8, 10, 12)
lty <- c(3, 1, 3, 1)
lwd <- c(1, 1, 2, 2)
mapContour(G$xg, G$yg, G$zg, lwd=lwd,
```

```
levels=levels, lty=lty)
## redraw the land to clean up contour lines
mapPolygon(coastlineWorldFine, col="gray")
legend("bottomright", lwd=lwd, lty=lty,
       legend=levels, bg="white", seg.len=4)
```

Detailed comparison (not shown here) suggests that the procedure represents the data reasonably well. Also, with computation time reduced from over a minute to a fraction of a second, the procedure is suitable for interactive analysis.

5.9.6 Differential Equations

5.9.6.1 Initial Value Problems

Consider slab-like, stress-driven flow on an f-plane, described by

$$\frac{du}{dt} - fv = F - \lambda u$$
$$\frac{dv}{dt} + fu = \quad - \lambda v$$
(5.19)

where u and v are velocity components in (say) the east and north directions, t is time, f is the Coriolis parameter, F is applied stress in the x direction, and a linear representation of bottom friction is used, with coefficient λ, assumed constant.[24] Given initial conditions, i.e. u and v at $t = 0$, these equations may be integrated with results as in Fig. 5.30.

The first step is to define the momentum equations, with

```
me <- function(t, y, parms=list(f=1e-4,
lambda=3e-5, F=1e-4))
{
    u <- y[1]
    v <- y[2]
    dudt <- parms$F + parms$f * v - parms$lambda * u
    dvdt <- -parms$f * u - parms$lambda * v
    list(c(dudt, dvdt))
}
```

Here, y holds u and v (i.e. the present state), and parms holds the parameters f, λ and F. The function returns a list containing du/dt and dv/dt. These coding patterns are explained in the documentation for lsoda in the deSolve package.

---

[24]Being linear, this system is easy to solve analytically. Even so, numerical approaches make it easy to predict the detailed response to complicated wind stress variation, etc., and permit easy extension to, e.g., nonlinear friction.

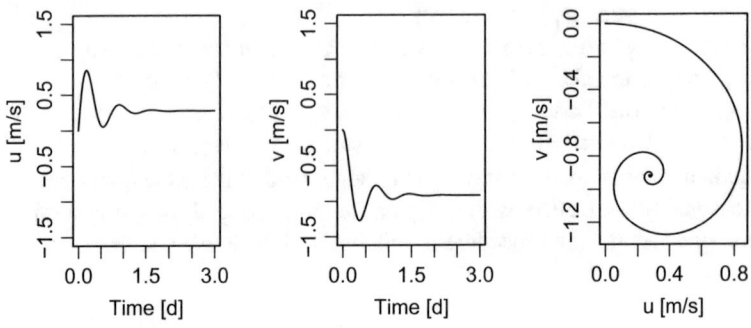

Fig. 5.30 Numerical estimate of solution of ordinary differential equations for wind-driven slab motion on a frictional f-plane

The next step is to set up initial conditions and define f, λ and F

```
IC <- c(0, 0)
parms <- list(f=1e-4, lambda=3e-5, F=1e-4)
```

after which it remains to define report times and integrate

```
t <- 86400 * 3 * (seq(0, 500) / 500) # 3 days
library(deSolve)
sol <- lsoda(IC, t, me, parms)
```

The results are plotted in Fig. 5.30 with

```
u <- sol[,2]
v <- sol[,3]
par(mfrow=c(1, 3), mar=c(4,4,1,1))
day <- t / 86400
plot(day, u, "l", xlab="Time [d]", ylab="u [m/s]",
     ylim=c(-1.5,1.5))
plot(day, v, "l", xlab="Time [d]", ylab="v [m/s]",
     ylim=c(-1.5,1.5))
plot(u, v, "l", xlab="u [m/s]",ylab="v [m/s]", asp=1)
```

Exercise 5.28 Construct a two-layer box model for temperature, with the top layer subjected to a sinusoidal heat flux. Assume the boxes to be of equal thickness, and devise a convection scheme to prevent temperature inversion. (See page 234 for a solution.)

Exercise 5.29 Develop a numerical solution to the convection problem formulated by Stommel (1961), $dy/dt = 1 - y - (y/\lambda)|Rx - y|$ with $dx/dt = \delta(1 - x) - (x/\lambda)|Rx - y|$, where x is dimensionless salinity, y is dimensionless temperature, t is dimensionless time, $\delta = 1/6$, $\lambda = 1/5$ and $R = 2$. Draw some traces to mimic Stommel's Figure 7, including one starting at $x = 0.55$, $y = 1$, which approaches a stable-spiral attractor. (See page 234 for a solution.)

5.9.6.2 Boundary Value Problems

Consider the Ekman spiral equations

$$-fv = A_v \frac{d^2u}{dz^2} \quad \text{and} \quad fu = A_v \frac{d^2v}{dz^2} \tag{5.20}$$

where u and v are the horizontal components of velocity, f is the Coriolis parameter, and A_v is a turbulent viscosity, assumed to be constant with respect to the vertical coordinate z. If a constant wind stress τ aligns with y, stress continuity yields surface boundary conditions

$$\frac{du}{dz} = 0 \quad \text{and} \quad \frac{dv}{dz} = \frac{\tau}{\rho A_v} \tag{5.21}$$

at $z = 0$. Velocity vanishes far below the surface, so $u \to 0$ and $v \to 0$ as $z \to -\infty$.

Nondimensionalization yields similar equations, but without f, A_v, τ and ρ. Converting second-order derivatives into coupled first-order derivatives, with shears denoted as S_x and S_y, the dynamical equations become

$$\frac{du}{dz} = S_x, \quad \frac{dv}{dz} = S_y, \quad \frac{dS_x}{dz} = -v, \quad \text{and} \quad \frac{dS_y}{dz} = u \tag{5.22}$$

with boundary conditions $S_x = 0$ and $S_y = 1$ at $z = 0$, along with $u \to 0$ and $v \to 0$ for $z \to -\infty$.

The first step to a solution is to define a function returning the derivatives, e.g. if y contains (u, v, S_x, S_y) then the following suffices

```
func <- function(z, y, parms)
{
    return(list(c(y[3], y[4], -y[2], y[1])))
}
```

Using bvptwp() from the bvpSolve package for this 2-point boundary value problem requires stating conditions at an initial point ($z \to -\infty$), i.e.

```
yini <- c(0, 0, NA, NA)
```

indicates that velocities vanish in the deep water. (No statement need be made about shears S_x and S_y there, so NA is used.) Velocities are left to be undetermined at the surface, but shears must be specified there

```
yend <- c(NA, NA, 0, 1)
```

A solution may now be found with bvptwp(), but as a practical matter it cannot cover the range $-\infty < z < 0$ but must instead cover a finite range, e.g. a solution for $-10 < z < 0$ may be found with

```
z <- seq(-10, 0, 0.1)
library(bvpSolve)
res <- bvptwp(yini=yini, yend=yend, x=z, func=func)
```

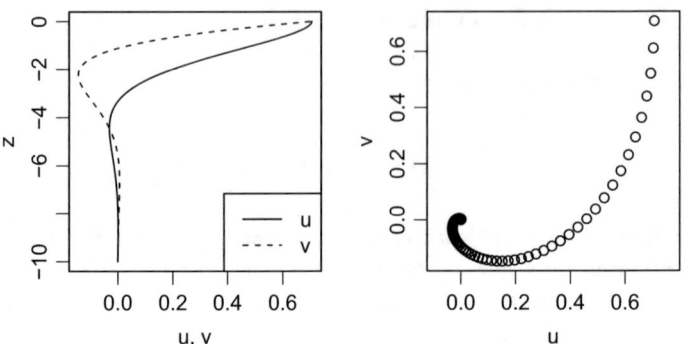

Fig. 5.31 Numerical solution to the Ekman spiral

after which res is a matrix of 5 columns, one for the independent coordinate z, and the others for u, v, S_x, and S_y.

The following displays the solution as Fig. 5.31, using both profiles and a hodograph

```
par(mfrow=c(1, 2))
plot(res[,2], res[,1], type="l",xlim=range(res[,2:3]),
     xlab="u, v", ylab="z")
legend("bottomright", lty=c("solid","dashed"),
        legend=c("u", "v"),bg="white")
lines(res[,3], res[,1], lty="dashed")
plot(res[,2], res[,3], type="p", asp=1,
xlab="u", ylab="v")
```

Note the spiralling form of the hodograph, hence the name of this situation.

5.9.6.3 Partial Differential Equations

In a seminal paper, Henry Stommel suggested that the variation of the Coriolis parameter with latitude provides an explanation for the narrowness and swiftness of currents such as the Gulf Stream (Stommel 1948). A nondimensional form of the relevant partial differential equation for the streamfunction ψ (from which velocities can be computed) is

$$\frac{\partial^2 \psi}{\partial x^2} + \frac{\partial^2 \psi}{\partial y^2} + \alpha \frac{\partial \psi}{\partial x} = \gamma \sin(\pi y) \tag{5.23}$$

where x and y are coordinates aligned to the east and north, the sinusoidal term with γ a constant describes the dependence of eastward wind stress on y, and α is proportional to df/dy, where f is the Coriolis parameter. The boundary condition

$\psi = 0$ (meaning no flow out of the domain) is applied at $x = 0$, $x = A$, $y = 0$ and $y = 1$, where A is the domain aspect ratio.

This system can be solved numerically using `tran.2D()` and `steady.2D()` from the ReacTran and rootSolve packages, loaded with

```
library(ReacTran)
library(rootSolve)
```

A grid with aspect ratio as in Stommel (1948) may be created with

```
n <- 100
xg <- seq(0, 1.6, length.out=n)
yg <- seq(0, 1, length.out=n)
dx <- diff(xg[1:2])
dy <- diff(yg[1:2])
```

It is necessary to set up a function for $\partial\psi/\partial t$ in a related dynamical system that has as its steady-state solution the ψ field sought here[25]:

```
de <- function(t, y, parms)
{
    pmat <- matrix(y, nrow=n, ncol=n)
    p <- tran.2D(pmat, D.x=1, D.y=1,
                 dx=dx, dy=dy, v.x=-parms$alpha,
                 C.x.up=0, C.x.down=0, C.y.up=0,
                 C.y.down=0)
    windForcing <- matrix(parms$gamma * sin(pi * yg),
                          nrow=n, ncol=n, byrow=TRUE)
    dP <- p$dC + windForcing
    return(list(as.vector(dP)))
}
```

The `y` argument is a vector of the solution, which is converted to a matrix so `tran.2D()` can calculate $\partial^2\psi/\partial x^2 + \partial^2\psi/\partial y^2 + \alpha\partial\psi/\partial x$.

After defining a matrix of (somewhat arbitrary) initial conditions

```
p0 <- matrix(0, nrow=n, ncol=n)
```

it remains to set up the details of a particular situation, here with `parms` chosen to mimic Figure 5 of Stommel (1948), and then to solve for a steady state and display the results as in Fig. 5.32,

```
p <- steady.2D(y=p0, func=de, parms=list(alpha=50,
               gamma=2e3), dimens=c(n, n), lrw=1e+6)
sol <- matrix(p$y, nrow=n, ncol=n)
contour(xg, yg, sol, labcex=1,
        xaxs="i", yaxs="i", asp=1,
        levels=seq(10, 40, 10))
```

---

[25]See Soetaert et al. (2010) and Chapter 3 of Soetaert and Herman (2009) for the mathematical concepts and framework of the R solution method.

Fig. 5.32 Numerical solution
of a nondimensional form of
the Stommel (1948) equations
for the streamfunction for
wind-driven ocean circulation

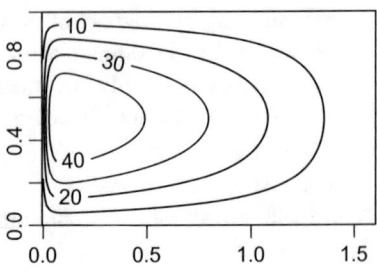

which is a clear display of the western intensification of circulation that was a key finding of the Stommel (1948) analysis.

5.9.7   *Optimization*

Numerical optimization of a function $f = f(x_1, x_1, \ldots)$ amounts to finding values of x_1, x_2, $\ldots$ for which the value of f is either a minimum or maximum in some neighbourhood. There are many ways to accomplish this (see, e.g., Nocedal and Wright 1999), and several are provided by R with the one-dimensional function `optimize()` and the more complicated multi-dimensional function `optim()`. Since `optimize()` has been treated already (e.g. in Sect. 2.3.11.4), the present section focuses on `optim()`.

The goal is an attempt to infer geographic location from sunrise and sunset times, which may be possible because of an intersection of the edges of the illuminated earth during the two events. (This fails at the equinoxes because then the edges of the illuminated half-spheres trace longitude lines.) Solar elevations may be computed with `sunAngle()` in the `oce` package, and combining this with `optim()` might permit location estimation based on sunrise and sunset times.

For example, consider Halifax, Nova Scotia, on Canada Day of the year 2017, when the sun rose at 5:33AM and fell at 9:03PM. These times, which are rounded to the nearest minute, may be represented in R with

```
library(lubridate)
tr <- with_tz(as.POSIXct("2017-07-01 05:33:00",
                         "America/Halifax"), "UTC")
ts <- with_tz(as.POSIXct("2017-07-01 21:03:00",
                         "America/Halifax"), "UTC")
```

where `with_tz()` from the `lubridate` package provides a convenient way to switch from the local timezone to UTC. Halifax is shown with a circle in Fig. 5.33, constructed with

```
LAT <- 44+38/60
LON <- -(63+35/60)
data(coastlineWorldFine, package="ocedata")
```

Fig. 5.33 Result of using
`optim()` to infer locations
from sunrise and sunset times

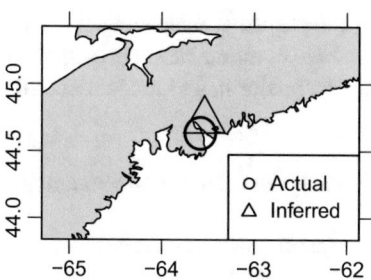

```
plot(coastlineWorldFine, span=250,
    clongitude=LON, clatitude=LAT)
points(LON, LAT, cex=3, lwd=3)
```

The question is whether the sunrise and sunset times, rounded to the nearest minute, can yield a location close to the actual value, at this time of year. A function giving the sum of squared solar angles above the horizon is

```
fn <- function(ll) {
    r <- sunAngle(tr,ll[1],ll[2],
    useRefraction=TRUE)$altitude
    s <- sunAngle(ts,ll[1],ll[2],
    useRefraction=TRUE)$altitude
    r^2 + s^2
}
```

With this, a solution is found and indicated in Fig. 5.33 with

```
o <- optim(c(0, 0), fn)
points(o$par[1], o$par[2], pch=2, cex=3, lwd=3)
legend("bottomright", pch=1:2, legend=c("Actual",
    "Inferred"))
```

The inferred location is within about 10 km of the actual value. (This example can be a good way to get children interested in the world around them, and even in history (see, e.g., Sobel 1995).)

Exercise 5.30 In her 1972 song "You're so vain," Carly Simon mentions flying to Nova Scotia to view a solar eclipse. Determine the time of that eclipse, assuming that it occurred on March 7, 1970. Use `optimize()` with the oce functions `moonAngle()` and `sunAngle()`. (See page 235 for a solution.)

5.9.8 Eigenanalysis Methods

Eigenanalysis is employed in many fields of study, and this enables advantageous cross-fertilization. For example, the Bretherton et al. (1992) discussion of ways to discover coupled patterns in climate data has clear analogies with oceanographic applications. R makes eigenanalysis easy, as will be illustrated here with sketches of three common tasks: principal component analysis, empirical orthogonal function decomposition, and modal decomposition.

5.9.8.1   Principal Component Analysis

Principal component analysis (PCA) is a common operation in oceanography. The method may be easiest to understand for (x, y) data, which can be illustrated with horizontal components of velocity in the adp dataset that comes with the oce package.

Out of general interest, a sensible first step may be to display the data as in the left panels of Fig. 5.34, with

```
data(adp, package="oce")
##'scale' for equal zero-centred scales on each panel
scale <- max(abs(adp[["v"]][,,1:2]), na.rm=TRUE)
plot(adp, which=1:2, drawTimeRange=FALSE,
     mar=c(2,3,1,1), zlim=scale*c(-1, 1))
```

revealing high similarity in patterns, which is not surprising in a tidal estuary. PCA provides a way to determine the dominant flow direction. The zone between 20 m and 30 m from the sensor will be analysed to avoid spurious data near the top of the water column. Velocity is in a three-dimensional matrix within adp, so the horizontal components in the focus zone may be extracted and plotted as in the right panel of Fig. 5.34 with

```
U <- adp[["v"]]
iDepth <- 20 <= adp[["distance"]] &
```

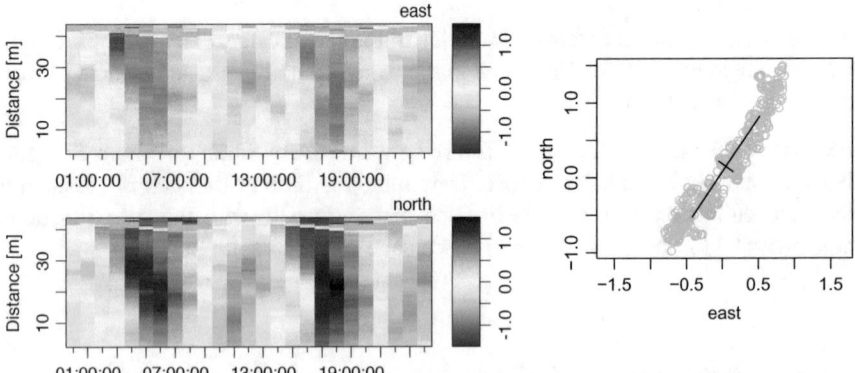

Fig. 5.34 Left: time and distance variation of the eastward and northward velocities in the adp dataset. The grey patches near the tops of the panels indicate poor data quality near the surface, revealing a tidal variation in water depth. Right: Principal component analysis of horizontal components of velocity measured 20 m to 30 m from the instrument

```
adp[["distance"]] <= 30
east <- as.vector(U[, iDepth, 1])
north <- as.vector(U[, iDepth, 2])
plot(east, north, asp=1, col="gray")
```

The PCA procedure is a simple matter of using `eigen()` for an eigenanalysis of a covariance matrix created with `cov()`, as follows:

```
C <- cov(data.frame(east, north), use="complete.obs")
e <- eigen(C)
```

after which `print(e)` reveals that the first eigenvalue is much larger than the second, indicating the thinness of the current ellipse. This can be shown by drawing a scaled principle axes cross in Fig. 5.34, with

```
S <- sqrt(e$values)
E <- e$vectors
me <- mean(east, na.rm=TRUE)
mn <- mean(north, na.rm=TRUE)
lines(me+c(-1,1)*S[1]*E[1,1],
mn+c(-1,1)*S[1]*E[2,1], lwd=3)
lines(me+c(-1,1)*S[2]*E[1,2],
mn+c(-1,1)*S[2]*E[2,2], lwd=3)
```

The angle of the major principal-component axis is

```
angle <- atan2(E[2,1], E[1,1]) * 180 / pi
```

i.e., 55.2° anticlockwise from east (see Fig. 5.10 for geographical context).

Although using `cov()` and `eigen()` is simple, and perhaps preferred by those experienced in linear algebra, the operations can be handled with `prcomp()`, which has the advantage of associated functions, notably `biplot()`.

5.9.8.2 Empirical Orthogonal Functions

Empirical orthogonal functions (EOFs) came into common use in meteorology in the 1980s (Wallace and Dickinson 1972; North et al. 1982; North 1984; Barnett and Hasselmann 1979). The basic idea is to form a basis set from the eigenvectors of a covariance matrix. The eigenvectors are often called "modes" although they need not bear any relationship to physical modes.

The `adp` dataset used in the previous section can also be used to illustrate EOF analysis. As noted before, the data for the top several meters of the water column are spurious, and so it makes sense to trim them with

```
adp2 <- subset(adp, distance < 38)
```

Now, an EOF decomposition of the depth dependence of the eastward velocity component is done with

```
u <- adp2[["v"]][,,1]
e <- eigen(cov(u))
```

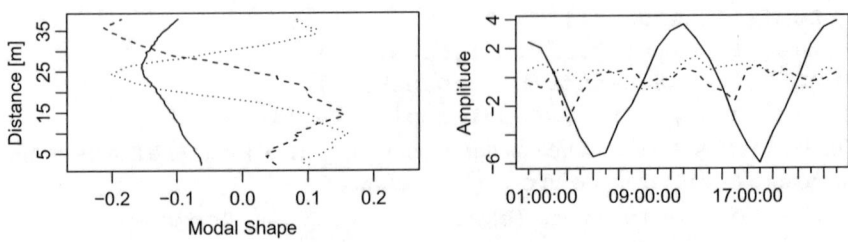

Fig. 5.35 Empirical orthogonal function analysis of the adp dataset. Left: shapes of modes 1, 2 and 3 (solid, dashed and dotted). Right: variation of mode amplitude with time

The eigenvectors give the modes and the eigenvalues give the variance explained by them. Computing

```
sum(e$values[1:3]) / sum(e$values)
[1] 0.9781308
```

shows that the first 3 modes hold most of the overall variance. These are isolated with

```
mode1 <- e$vectors[,1]
mode2 <- e$vectors[,2]
mode3 <- e$vectors[,3]
```

and plotted in the left panel of Fig. 5.35 with

```
distance <- adp2[["distance"]]
plot(mode1, distance, type="l", xlim=c(-1,1)/4,
     xlab="Modal Shape", ylab="Distance [m]")
lines(mode2, distance, lty=2)
lines(mode3, distance, lty=3)
```

The temporal variation of modal amplitude is calculated by projecting the modes onto the data

```
a1 <- (u %*% e$vectors[,1])[,1]
a2 <- (u %*% e$vectors[,2])[,1]
a3 <- (u %*% e$vectors[,3])[,1]
```

The mode amplitudes are plotted in the right panel of Fig. 5.35 with

```
time <- adp2[["time"]]
oce.plot.ts(time, a1, ylab="Amplitude",
drawTimeRange=FALSE)
lines(time, a2, lty=2)
lines(time, a3, lty=3)
```

A common use of empirical orthogonal function analysis is to construct simplified models of observed variability. For example, a simplified model with the first 3 modes of the data under consideration can be constructed and plotted in Fig. 5.36 with

```
U <- t(outer(mode1,a1) + outer(mode2,a2)
+ outer(mode3,a3))
```

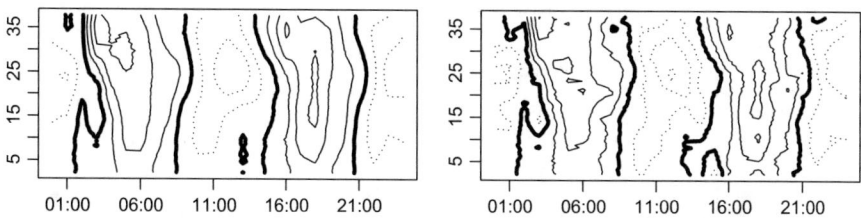

Fig. 5.36 Two-mode EOF reconstruction of `adp` dataset (left) compared with data (right). Contours at interval 0.25 m/s, with negatives dotted and zero highlighted

```
contour(time, distance, U,
levels=seq(-2, -0.25, 0.25),
          lty=3, drawlabels=FALSE)
```
plus other `contour()` calls distinguishing zero and positive values by line width and type. The right panel is constructed similarly, but for the actual data. Comparison reveals that these three modes capture the main patterns, which is as expected, given the fraction of variance they explain.

Exercise 5.31 Use `svd()` to apply the singular value decomposition to the `adp` dataset. (See page 236 for a solution.)

5.9.8.3   Dynamical (Internal Wave) Modes

Under certain conditions, the vertical isopycnal displacement $\phi = \phi(z)$ associated with internal waves may be described by

$$\frac{1}{N^2}\frac{d^2\phi}{dz^2} + \frac{1}{C^2}\phi = 0 \tag{5.24}$$

where $N^2 = -(g/\rho)\partial\rho/\partial z$ is the square of the buoyancy frequency and C is a scalar that yields the horizontal propagation speed ("celerity") of a modal solution (see, e.g., Cushman-Roisin and Beckers (2011, Section 13.4) or Gill (1982, Section 6.11)). Assuming no displacement through the mean surface at $z = 0$ and at the (assumed flat) bottom at $z = -H$, the boundary conditions are $\phi = 0$ at $z = 0$ and $z = -H$.

Equation (5.24) may be expressed in finite-difference form as

$$\frac{1}{N_i^2\Delta z^2}(\phi_{i-1} - 2\phi_i + \phi_{i+1}) + \frac{1}{C^2}\phi_i = 0 \tag{5.25}$$

where Δz is the distance between the levels and i indexes those levels. In matrix notation, introducing $d_i = 1/(N_i\Delta z)^2$ for brevity, this is

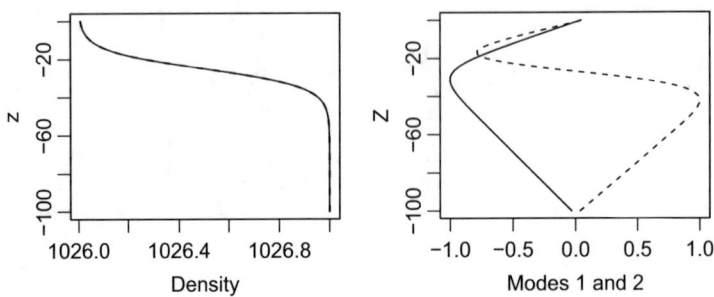

Fig. 5.37 A density profile and its first two internal-wave modes

$$
\begin{bmatrix}
0 & 0 & 0 & 0 & 0 & \cdots \\
d_2 & -2d_2 & d_2 & 0 & 0 & \cdots \\
0 & d_3 & -2d_3 & d_3 & 0 & \cdots \\
\vdots & \vdots & \vdots & \vdots & \vdots & \ddots
\end{bmatrix}
\begin{bmatrix}
\phi_1 \\ \phi_2 \\ \phi_3 \\ \vdots
\end{bmatrix}
+ \frac{1}{C^2}
\begin{bmatrix}
\phi_1 \\ \phi_2 \\ \phi_3 \\ \vdots
\end{bmatrix}
= 0
\qquad (5.26)
$$

This equation is of the form $D\phi = \lambda\phi$, in which D is a dynamics matrix, ϕ is an eigenvector and $\lambda = -C^{-2}$ is an eigenvalue. A solution may be found with `eigen()`. For example, a hyperbolic-tangent stratification is defined and plotted in the left panel of Fig. 5.37 with

```
z <- seq(-100, 0, 1)
rho <- 1026.5 - 0.5*tanh((z+25)/10)
par(mfrow=c(1, 2))
plot(rho, z, type="l", xlab="Density")
```

For such a smooth profile, N^2 could be computed by first difference, but to illustrate a useful step for noisy data, `smooth.spline()` may be can be used before first-differencing

```
spline <- smooth.spline(z, rho)
Rho <- predict(spline)
lines(Rho$y, Rho$x, lty=2, lwd=2)
```

The two lines coincide in Fig. 5.37, as expected. However, with more ragged profiles, or simply for greater control, an analyst might consider providing `smooth.spline()` with more arguments to control the smoothing, e.g. df to set the number of degrees of freedom. Once an acceptable model profile is established, the required derivatives can be found with the `predict()` function for a smoothing spline, e.g.

```
DZ <- 2
Z <- seq(min(z), max(z), DZ)
N2 <- -9.8 / mean(rho) * predict(spline, Z, deriv=1)$y
```

where Z is the desired computation grid. Now, D may be constructed with

```
n <- length(N2)
d <- 1 / (N2 * DZ^2)
D <- matrix(0, nrow=n, ncol=n)
for (r in 1:n) {
    D[r, max(1, r-1)] <- d[r]
    D[r, r] <- -2*d[r]
    D[r, min(n, r+1)] <- d[r]
}
```

where `max()` and `min()` prevent overstepping the matrix limits. The actual decomposition into dynamical modes is simple, with

```
e <- eigen(D)
```

As in the previous section, e is now a list holding eigenvalues and eigenvectors. Physical interpretation is simplified if the eigenvectors are ordered by speed (ignoring eigenvalues of nonphysical sign), and if the mode shapes are normalized. The results of

```
C <- ifelse(e$values < 0, sqrt(-1/e$values), 0)
o <- order(C, decreasing=TRUE)
C <- C[o]
modes <- e$vectors[,o]
for (i in 1:dim(modes)[2])
    modes[,i] <- modes[,i] / max(abs(modes[,i]))
plot(modes[,1], Z, xlim=c(-1,1), type="l",
        xlab="Modes 1 and 2")
lines(modes[,2], Z, lty="dashed")
```

are shown in Fig. 5.37. Note that the first of these modes has one lobe, that the second has two, and that each shape might be seen as a stretched version of the corresponding sinusoidal mode resulting from constant stratification.

Exercise 5.32 Explore the accuracy of this method by adjusting Δz with the case of constant $N = 0.01\mathrm{s}^{-1}$. (See page 237 for a solution.)

5.9.9 Neural Networks and Machine Learning

A neural network is a pattern-recognition technique that is modelled loosely on assemblages of biological neurons. The basic idea is that signals pass between neurons via nonlinear processes, and that tailoring those processes can create a system that maps certain input patterns to certain output values. For example, an input signal might be a listing of water properties at each level within a CTD profile, and the output might be a Boolean value indicating whether the CTD had passed through an overturning eddy (see, e.g., Galbraith and Kelley 1996). Importantly, the analyst is not required to devise a conventional algorithm to identify the pattern of interest. The procedure demands only a decision about network geometry and then the provision of a training set with adequate sampling of situations that may come up in applications.

Neural networks can be helpful in tasks ranging from plankton identification based on flow cytometry (Boddy et al. 1994, 2000) and digital holographic microscopy (Missan et al. 2018) to El Niño prediction based on large-scale atmospheric and oceanic properties (Tangang et al. 1998; Wu et al. 2006). An early introduction to the use of neural networks in meteorology and physical oceanography was given by Hsieh and Tang (1998), and the later textbook by Hsieh (2009) provides updates and further details.

Caution should be exercised in approaching the literature on neural networks, given an occasional tendency to exaggerate the power of the method (see, e.g., Hutson 2018). This tendency has been remarked upon since the early days of research in this field; see Ripley (1994) and Ripley (1996) for wide-ranging insights on this and related matters, especially in the statistical context.

There are several ways to perform neural network analysis in R. Bergmeir and Benítez (2012) provide a good entry to the topic, as well as a helpful comparison of some of the popular packages. Here, the nnet package will be used for illustration, partly because it is discussed in some detail within the important textbook by Venables and Ripley (1999).

Consider the `drag` dataset, which contains drag-coefficient data digitized from Garratt (1977, Figure 3). As discussed in Sect. 2.3.7, the contents consist of measurements of drag coefficient C_D, at different wind speeds U, made with two different methods. A neural network expressing Cd as a function of U and method may be constructed with

```
data(drag, package="ocedata")
library(nnet)
n <- nnet(1000*Cd~U+method, data=drag,
          size=2, linout=TRUE, decay=1e-2,
          maxit=1000)
```

The first nnet() argument is a formula expressing the relationship (for variables within the second argument), which in this case scales C_D so it will be of order 1, to prevent the optimizing function used by nnet() from having difficulties finding a solution. Note that the formula should not be read as a regression formula; it indicates merely that C_D depends in some way on U and *method*. The size value controls the network size. The linout argument specifies that the output should be linear, as opposed to logistic, the latter being better for classification problems. Adjusting decay can prevent overfitting. The number of iterations is set by maxit. Other nnet() arguments are explained in its documentation and in Sections 9.4 and 11.6 of Venables and Ripley (1999).

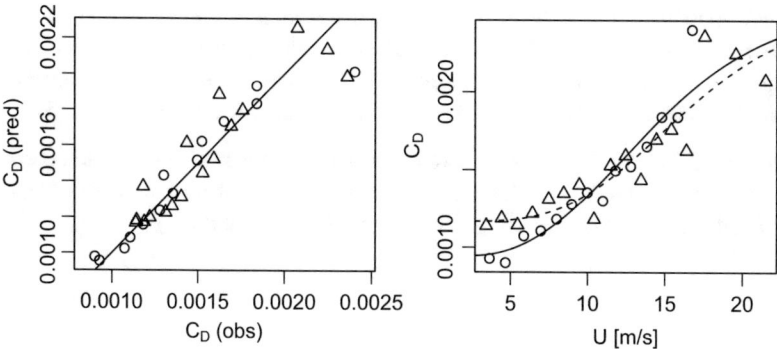

Fig. 5.38 Neural network analysis of the drag dataset. Left: observed and predicted C_D values, with circles for the profile method and triangles for the eddy method. Right: dependence on wind speed, with solid and dashed lines for the profile and eddy predictions

The C_D prediction is compared with observations in the left panel of Fig. 5.38, created with

```
pch <- ifelse(drag$method=="eddy", 2, 1)
plot(drag$Cd, 1e-3*predict(n), asp=1, pch=pch,
     xlab=expression(C[D]*" (obs)"),
     ylab=expression(C[D]*" (pred)"))
abline(0, 1)
```

The right panel of Fig. 5.38, created with

```
plot(drag$U, drag$Cd, pch=pch,
     xlab="U [m/s]", ylab=expression(C[D]))
U <- seq(0, 30, 0.2)
lines(U, 1e-3*predict(n, list(U=U,
method="profile")))
lines(U, 1e-3*predict(n, list(U=U,
method="eddy")), lty=2)
```

illustrates that the predictions run smoothly through the experimental data, distinguishing between the profile and eddy inference methods.

In this and other simple prediction cases, a single-layer network such as that made available with nnet() may be sufficient. However, more challenging pattern-recognition problems can benefit from multiple layers, as are provided by neuralnet and other systems in R.

Since machine learning is an area of active research, a good starting point for R methodologies is the CRAN task view dedicated to the topic,[26] because it is updated frequently. Of particular interest for challenging pattern recognition problems is the advent of deep learning techniques, a topic for which (LeCun et al. 2015)

---

[26]https://cran.r-project.org/web/views/MachineLearning.html.

provide an excellent introduction. In that connection, it is worth pointing out that the `tensorflow` package links R with the TensorFlow system,[27] a machine learning library that has seen wide application in both scientific research and commercial products, since Google released it with an open-source licence in late 2015.

Exercise 5.33 Create a neural network to describe the variation of sound speed with pressure, for station 103 of the `section` dataset, exploring the effect of variable transformation on reproducibility. (See page 237 for a solution.)

---

[27] https://www.tensorflow.org.

Chapter 6
Solutions

Abstract This chapter presents potential solutions to the exercises presented in the previous chapters, along with additional discussion of related issues. The exercises range widely in complexity, with even the most basic being worthy of some attention. For example, readers who take the first exercise seriously will learn not just how to construct simple linegraphs, but also how to read R documentation, and that is a skill that can pay off in more sophisticated applications such as mapping the classic Endeavour cruise, as shown in Fig. 6.1.

6.1 Exercises in Chap. 2

Exercise 2.1 on page 7. Type `help(plot)` in a console, and use the results to see how to draw a line graph instead of a scatter plot.

Solution. The result of typing the indicated text in an R console is a "help" page. Reasoning that the solution is likely to lie in an argument to `plot()`, a clever reader will skip to the documentation section dealing with arguments, and see the solution quickly:

```
plot(xy[,1], xy[,2], type="l")
```

gives a line graph. Readers should also try other `type` options: `"s"` for a staircase pattern, `"h"` for a staircase with vertical lines (as in a histogram), `"b"` for both points and lines, the latter drawn with spaces adjacent to the points, and `"o"` for over-plotting of points and lines.

Exercise 2.2 on page 8. Consult the documentation for `read.table()`, to see how to indicate that the first line of the file contains a line with the names of the columns.

Solution. Typing `help(read.table)` produces results that are a bit daunting, but simpler than they may seem at first. The "Usage" section lists `read.table()` and some other functions that are similar enough to be grouped together. Focussing just on `read.table()`, one sees that the arguments are named `file`, `header`, `sep` and so on. As in the previous exercise, these are explained in the "Arguments"

© Springer Science+Business Media, LLC, part of Springer Nature 2018
D. E. Kelley, *Oceanographic Analysis with R*,
https://doi.org/10.1007/978-1-4939-8844-0_6

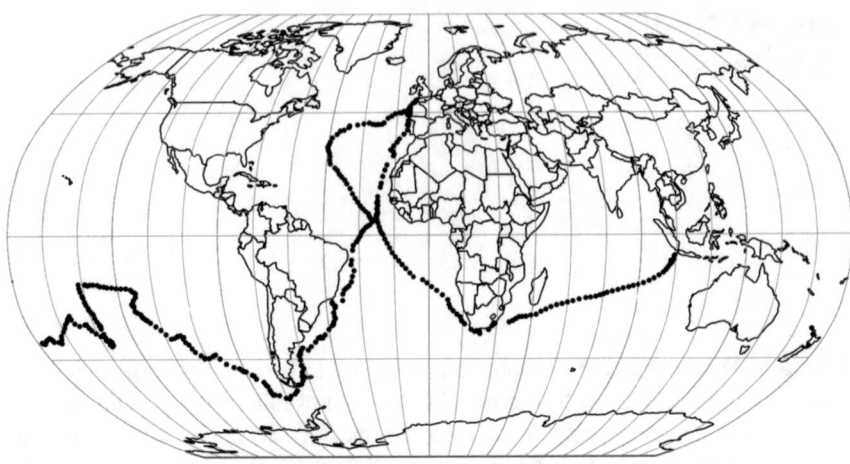

Fig. 6.1 H.M.S. Endeavour cruise, 1768–1771, shown in Robinson projection. (See Page 210 for the R code that creates this diagram)

section. Clearly, `file` can be a file name, so there is no need (yet) to study the rest of the information on that argument. The next argument is `header`, which is "a logical value indicating whether the file contains the names of the variables on its first line". This reveals that the solution lies in adding one more argument:

```
xy <- read.table("xy.dat", header=TRUE)
```

Exercise 2.3 on page 8. Use the `mfrow` argument of `par()` to draw multi-panel plots in R, emulating the Matlab `subplot` command.
Solution. A great deal of information is provided by `help(par)`, but readers who did the last two exercises will know that R documentation follows a fixed format that makes skimming easy, revealing that, e.g., `par(mfrow=c(3,2))` gets R ready to plot a six-panel plot with 3 rows and 2 columns. What happens is that the next six plots are drawn on this grid, with the top row being filled up first. (By contrast, `par(mfcol=c(3,2))` fills the grid by columns.) Readers who are familiar with Matlab should note that no further commands are needed to specify where to place the plots, which may be a plus, but that finished panels cannot be revised, which is certainly a minus.

Exercise 2.4 on page 8. Use `outer()` to emulate the Matlab function `meshgrid`.
Solution. Near the start of the documentation for `outer()`, one reads "The outer product of the arrays X and Y is the array A with dimension `c(dim(X)`, `dim(Y))` where element `A[c(arrayindex.x, arrayindex.y)]` = `FUN(X[arrayindex.x], Y[arrayindex.y], ...)`". This deserves some thought. The documentation for `dim()` reveals that it returns the dimension of its argument. So, given vector values of X and Y, A is a matrix with as many columns as items in X. A similar statement holds for the rows. The function FUN is provided by the user. The use of "dummy" variables `arrayindex.x` and `arrayindex.y`

indicates that FUN will be called sequentially, for all combinations of X and Y. For example, the Matlab code

```
x = 0:2:100;
y = 0:2:100;
[X, Y] = meshgrid(x, y);
z = exp(-((X-50).^2 + (Y-50).^2)/100);
```

defines a grid for x and one for y, then creates a peaked function centred at the midpoints of x and y. In R, vectors can be built with seq()

```
x <- seq(0, 100, 2)
y <- seq(0, 100, 2)
```

and outer() can build the desired matrix, e.g.

```
f <- function(x,y) {
    distanceSquared <- (x - 50)^2 + (y - 50)^2
    radiusSquared <- 10^2
    return(exp(- distanceSquared / radiusSquared))
}
z <- outer(x, y, f)
```

This shows an important difference between R and Matlab notation. In Matlab, a dot is used as a prefix for operators to be applied element by element (e.g. "." precedes "^" in the example above). In R, it is the matrix operations that get syntax decorations, e.g. %*% is an inner product.

Exercise 2.5 on page 12. Use help.find() to find an R package that accesses the www.geonames.org website, and thus locate Halifax, Nova Scotia.
Solution. Using

```
help.search("geonames")
```

reveals the geonames package. Its documentation suggests that GNsearch() provides a solution, viz.

```
library(geonames)
GNsearch(q="halifax canada", maxRows=1)
```

although it must be noted that this will not work without first registering as a user on the geonames website.[1]

Exercise 2.6 on page 14. Use cumsum() to monitor the convergence of the Taylor series for exp().
Solution. This is a simple matter of replacing sum() with cumsum(), and using exp() for the comparison, e.g.

```
cumsum(0.1^(0:4) / factorial(0:4)) / exp(0.1)
 [1] 0.9048374 0.9953212 0.9998453 0.9999962 0.9999999
```

illustrates quick convergence with small $|x|$.

---

[1] http://www.geonames.org.

Exercise 2.7 on page 17. Use == to find your computer's precision, i.e. the smallest resolvable difference between floating-point values.

Solution. The fact that

```
(1 + 1e-20) == 1
```

returns TRUE suggests that the precision is coarser than 10^{-20}, while a similar test shows it is finer than 10^{-10}. Rather than continue with trial values, it makes sense to construct a function that changes sign as the logical expression changes, and to use uniroot() to find the zero of this function; thus,

```
fcn <- function(x) ifelse((1+10^x)==1, -1, 1)
log10(10^uniroot(fcn, lower=-20, upper=-10)$root)
| [1] -15.95451
```

shows the machine has precision of order 10^{-16}, as expected for a 64-bit CPU using IEEE-754 double-precision arithmetic (IEEE Computer Society 2008). See the next exercise for practical implications.

Exercise 2.8 on page 17. Explain why all.equal() is good way to compare floating-point values.

Solution. The accuracy of real-world calculation tends to be much worse than machine precision. This is why the comparison function all.equal() uses the square root of machine precision as the default for its tolerance argument. On the author's computer, that default is 1.5×10^{-8}, but the value could be different on different machines. This tailoring to the computer yields important code portability. Another strength of all.equal() is the convenience afforded by its specialized versions for different data types. Serious analysts use all.equal() liberally and take the time to study its documentation.

Exercise 2.9 on page 19. A directory contains Biosonics echosounder files, with names indicating start times, with four digits for year, two for month and two for day, followed by an underline and then two digits for hour, two for minute, and two for second, ending with .dt4. Use grep() to isolate data starting between 1100 h and 1500 h on June 28th, 2008.

Solution. A solution is

```
f <- list.files(".")
files <- f[grep("^20080628_1[1234].*dt4$", f)]
```

Exercise 2.10 on page 25. Use floor() to select even integers from a vector.

Solution. We may use floor() and a multiplication-division pair to test for even numbers, as suggested

```
v <- 1:10
v[2*floor(v/2) == v]
| [1]   2   4   6   8 10
```

while another way is to use %% for modulo division

```
v[v%%2 == 0]
| [1]   2   4   6   8 10
```

Exercise 2.11 on page 28. Write a function to find the indices of the maximal value of a matrix.

Solution. A solution is

```
which.max2 <- function(m)
{
    ij <- which.max(m)
    ni <- dim(m)[1]
    j <- ceiling(ij / ni)
    i <- ij - (j - 1) * ni
    list(i=i, j=j)
}
```

and a demonstration with a coarse bathymetry file is

```
data(topoWorld, package="oce")
w <- which.max2(-topoWorld[["z"]])
topoWorld[["longitude"]][w$i] # near Challenger Deep
[1] 144.5
topoWorld[["latitude"]][w$j]
[1] 12
```

Exercise 2.12 on page 28. Show how to access a list within a list.

Solution. A sample list, with information on two ocean weather stations, is

```
ows <- list(name=c("Bravo", "Papa"),
            location=list(latitude=c(57, 50),
                          longitude=c(-50, -150)))
```

Suppose the task is to change longitude from degrees east to degrees west. In the dollar-sign notation, this can be done with

```
ows$location$longitude <- -ows$location$longitude
```

or with

```
ows[[2]][[2]] <- -ows[[2]][[2]]
```

in bracket notation, or with

```
ows[[2]]$longitude <- -ows[[2]]$longitude
```

in combined notation; see help("["].

Exercise 2.13 on page 31. Use factor() and split() to identify the months in which the Keeling CO_2 signal rises and falls.

Solution. R provides a dataset of Mauna Loa CO_2 concentration,[2] and

```
data(co2)
t <- time(co2)
dco2dt <- diff(co2) / diff(t)
## Shorten t and co2 to match length of dco2dt
```

---

[2]More detailed and up-to-date measurements of Manua Loa CO_2 concentration are provided at the Scripps website https://scripps.ucsd.edu/programs/keelingcurve/.

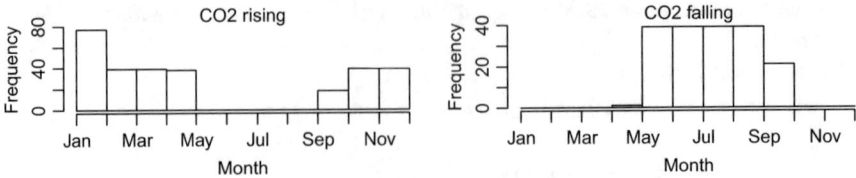

Fig. 6.2 Months in which Mauna Loa co2 rises and falls. (Exercise 2.13)

```
t <- t[-1]
co2 <- co2[-1]
```
isolates time and a first-difference estimate of $d\,CO_2/dt$. A factor identifying rising and falling concentration is calculated with
```
fac <- factor(dco2dt > 0, levels=c(TRUE, FALSE),
              labels=c("rising", "falling"))
```
after which the monthly histograms in Fig. 6.2 may be created with
```
month <- 1 + round(12*t%%1)# %% gets fraction of year
d <- data.frame(t, co2, month)
ds <- split(d, fac)
hist(ds$rising$month, breaks=1:12,
     xlab="Month", main="", axes=FALSE)
## Label by month name instead of number
axis(1, at=1:12, label=format(ISOdate(2018,1:12,1),
     "%b"))
axis(2)
mtext("CO2 rising", side=3)
```
and similar for ds$falling. The transition indicates peak concentration in May, as revealed in the seminal paper by Keeling (1960), just a few years into his important measurement program.

Exercise 2.14 on page 32. Construct a data frame with column x containing numbers from 0 to 2π, and y containing $\sin x$.
Solution.
```
x <- seq(0, 2*pi, length.out=100)
y <- sin(x)
df <- data.frame(x=x, y=y)
```

Exercise 2.15 on page 32. Append volume to the oceans dataset from the ocedata package.
Solution. A solution follows; note use of names()
```
data(oceans, package="ocedata")
names(oceans) # note that 'Volume' is not there
oceans$Volume <- oceans$Area * oceans$AvgDepth
```

Exercise 2.16 on page 32. Suppose a data frame contains CTD data for a series of stations, with columns for salinity, temperature, pressure, and station ID. Use `split()` and `factor()` to create a list with one element per station.

Solution. To illustrate, construct some fake data in two stations, one with two data levels, the other with three.

```
d <- data.frame(stn=c(1, 1, 2, 2, 2),
                S=c(33.1, 33.2, 35.1, 35.2, 35.3),
                T=c(10.1, 10.2, 15.1, 15.2, 15.3),
                p=c(1, 2, 1, 2, 3))
```

Now, use `factor()` and `split()` to create a list that will contain two items

```
stations <- split(d, factor(d$stn))
```

Exercise 2.17 on page 35. Devise a function using `ifelse` that returns the tangential velocity in a Rankine vortex.

Solution. This is defined by $r\Gamma/(2\pi R^2)$ for $r \leq R$ and $\Gamma/(2\pi r)$ otherwise:

```
rankine <- function(r, R, Gamma)
    ifelse(r < R, Gamma*r/(2*pi*R^2), Gamma/(2*pi*r))
```

Exercise 2.18 on page 36. Use `uniroot()` and `coriolis()` from the oce package, to find the critical latitude at which the Coriolis parameter f matches the M2 tidal frequency (12.4206 h period).

Solution. We must find the root ϕ_0 of the function $f(\phi) - \omega$, where $f = f(\phi)$ expresses the dependence of the Coriolis parameter on latitude ϕ, i.e.

```
uniroot(f=function(lat) (2*pi/12.4206/3600)
          - coriolis(lat), lower=0, upper=90)$root
[1] 74.47185
```

As an aside, the periods of common tidal frequencies can be found with the dataset named `tidedata` in the oce package, e.g.

```
data(tidedata, package="oce")
1 / subset(tidedata$const, "M2"==name)$freq
[1] 12.4206
```

Exercise 2.19 on page 36. Use `uniroot()` to create a function that calculates linear gravity wave speed as a function of period.

Solution. Linear theory yields an implicit equation for phase speed c_p in terms of period τ, depth H and gravitational acceleration g

$$0 = c_p - \frac{g\tau}{2\pi} \tanh \frac{2\pi H}{\tau c_p} \tag{6.1}$$

so that finding the root using `uniroot()` yields the desired function

```
cp <- function(tau, H, g=9.8)
    uniroot(function(x) x - g*tau/(2*pi)*tanh(2*pi*H/
            (tau*x)), interval=c(0, 100))$root
```

Exercise 2.20 on page 37. Create a function closure for individualized calibration of Seabird thermistors.

Solution. These thermistors are connected to systems that produce a signal of frequency that varies inversely with electrical resistance.[3] Temperature T (°C) is assumed to be related to frequency f (Hz) by a Steinhart-Hart equation

$$T = -273.15 + 1/[g + h\ln(f_0/f) + i\ln^2(f_0/f) + j\ln^3(f_0/f)] \qquad (6.2)$$

where $f_0 = 1000$ Hz is a constant, and g, h, i and j are calibration coefficients for individual thermistors. Expressing this as

```
Tcal <- function(g, h, i, j, f0=1000) {
    function(f) {
        ff <- log(f0 / f)
        -273.15 + 1 / (g + h*ff + i*ff^2 + j*ff^3)
    }
}
```

i.e., as a function that returns another function, yields a way to create customized calibrations, based on the particular values of g, h, i and j for individual thermistors, e.g.

```
T2132 <- Tcal(g=4.12744629e-3, h=6.26321187e-4,
              i=2.05376982e-5, j=2.13741203e-6)
```

creates a function for inferring temperature from Seabird thermistor serial number 2132, mentioned in the footnote, e.g. `T2132(3180.886)` yields 18.6605 °C (cf. calibration-bath temperature 18.6607 °C). Using a function closure means that the coefficients appear just once in the code, reducing the burden of keeping track of which set of coefficients to use for a given thermistor under analysis.

Exercise 2.21 on page 40. Write a loop that displays the values of items in the current workspace, using `ls()` and `get()`.

Solution. A `for` loop is a good way to handle this.[4] Note the use of an odd name (`.anItem`) to avoid printing the index of the loop.

```
for (.anItem in ls()) {
    if (.anItem != ".anItem") {
        cat(.anItem, "\n")
        print(get(.anItem))
        cat("\n")
    }
}
```

Exercise 2.22 on page 41. Extract velocity from the `oce` dataset adp, and plot distance-averaged beam-1 velocity versus time.

---

[3] See http://www.seabird.com/sbe3plus-ctd-temperature-sensor for more on Seabird thermistors and their calibration.

[4] A function to do this would need to use `ls(envir=parent.frame())` to access variables in the calling environment.

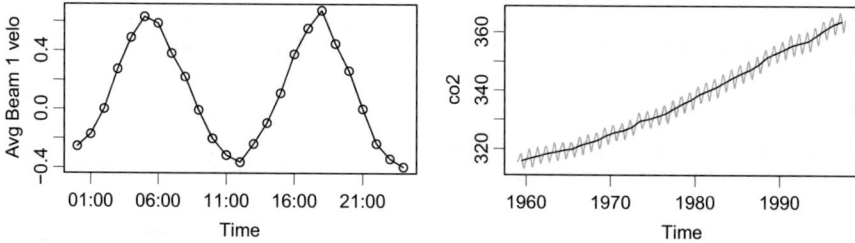

Fig. 6.3 Results from Exercises 2.22 and 2.23

Solution. Some exploration with `str()` reveals the names of the relevant quantities, which may be extracted

```
data(adp, package="oce")
t <- adp[["time"]]
v <- adp[["v"]]
```

Then, after checking dimensional matchup with

```
length(t)
```
```
[1] 25
```
```
dim(v)
```
```
[1] 25 84   4
```

one may write a solution as in Fig. 6.3 (left) with

```
v1 <- apply(v[,,1], 1, mean, na.rm=TRUE)
plot(t, v1, type="o", xlab="Time",
ylab="Avg Beam 1 velo")
```

Exercise 2.23 on page 41. Calculate and plot yearly average CO_2 data, using `lapply()`.

Solution. It is convenient to use the built-in `co2` dataset. This is stored as time-series object, so time may be extracted with

```
data(co2)
t <- time(co2)
```

which ranges from 1959 to 1997.917. Following a procedure similar to that of Exercise 2.13, a factor for yearly subdivision may be created with

```
tt <- 1959:1998
fac <- cut(t, breaks=tt)
```

allowing yearly averages to be constructed with

```
co2l <- split(co2, fac)
cc <- as.numeric(lapply(co2l, mean))
```

and results displayed as in Fig. 6.3 (right) with

```
plot(co2, type="l", col="darkgray")
lines(tt[-1]-0.5, cc)
```

Exercise 2.24 on page 41. Use a function in the `plyr` package to find minima and maxima of the data stored in `ctd[["data"]]`, a CTD station provided by the `oce` package.

Solution. First, use

```
data(ctd, package="oce")
class(ctd[["data"]])
[1] "list"
```

to load the CTD object and see that the item in question is a list. The next step is to find the relevant `plyr` function. This is easy, because `plyr` employs a naming scheme in which the first letter indicates the type of input (here, "l" for a list) and the second the type of desired output (here, d for a data frame). The second argument to be provided to `ldply()` is a function that returns a data frame, so a solution is

```
library(plyr)
ldply(ctd[["data"]],
        function(x) data.frame(min=min(x), max=max(x)))
```

where the 8 lines of output are omitted for brevity. Note that using `summary(ctd)` provides more extensive information about the dataset, including important metadata.

Exercise 2.25 on page 43. Reproduce Fig. 2.7 with axes labelled in geographical notation.

Solution. With `lon`, `lat` and `topo2` as in the text, use, e.g.

```
contour(lon, lat, topo2, drawlabels=FALSE, levels=c(0,-5000),
        col=c("black", "gray"), xaxs="i", yaxs="i", asp=1,
        axes=FALSE, xlim=c(-180,180), ylim=c(-90,90)) # new
box()
axis(1, at=c(-180, 0, 180), labels=c("180W", "0", "180E"))
axis(2, at=c(-90, 0, 90), labels=c("90S", "0", "90N"))
```

Exercise 2.26 on page 44. Devise a wrapper function to handle reversed x or y values in contouring.

Solution. The `order()` function returns the indices of a vector in an order that makes the values monotonic, so we may write

```
contour2 <- function(x, y, z)
{
    ox <- order(x)
    oy <- order(y)
    contour(x[ox], y[oy], z[ox, oy])
}
```

Note that this displays axes with increasing values. If this is not desired, e.g. if the y axis represents pressure, and it is desired to have high pressure at the bottom of the graph, use `order(y, decreasing=TRUE)`, or consider using `oceContour()`.

Exercise 2.27 on page 44. Contour the formula for wind-chill temperature $13.12 + 0.6215T - 11.37U^{0.16} + 0.3965TU^{0.16}$, as a function of air temperature, T, and wind speed, U.

Solution. This formula is suggested[5] for $T < 0\,°C$ and $U > 5\,km/h$, so a relevant graph (not shown) can be produced with

```
T <- seq(-20, 10, 1)
U <- seq(5, 50, 1)
chill <- function(T, U)
    13.12 + 0.6215 * T - 11.37 * U^0.16 + 0.3965 *
    T * U^0.16
contour(T, U, outer(T, U, chill))
```

Exercise 2.28 on page 50. Use the `panels` argument to draw the panels as density diagrams, using `smoothScatter()`.

Solution. This is not as simple as writing `panel=smoothScatter`, because `smoothScatter()` creates a new plot, and `panel` must be a function that adds to an existing plot. The solution is to create a new function that adds a smooth-scatter diagram to an existing plot, e.g.

```
pairs(d, panel=function(...) smoothScatter(..., add=TRUE))
```

Exercise 2.29 on page 55. The Rink Ratz® hockey card game has a 69-card deck with 2 desirable "miraculous save" cards. At the start of the game, 5 cards are discarded without being examined. What is the probability that there will be exactly 1 miraculous save card left in the deck?

Solution. There will be exactly one miraculous save (MS) card left if the discarded pile contains 1 MS card and 4 non-MS cards. The number of ways to pick 1 of the 2 MS cards is $^2C_1$, and the number of ways to pick 4 of the 67 non-MS cards is $^{67}C_4$. These are independent events, so the number of ways to get both is the product $^2C_1\,^{67}C_4$. The total number of ways to pick 5 cards from the deck of 69 is $^{69}C_5$, so the desired probability is

```
choose(2, 1) * choose(67, 4) / choose(69, 5)
[1] 0.1364024
```

so it can be expected to occur about once every 7 games.

Exercise 2.30 on page 56. Construct a graph comparing the normal distribution with the t distribution with 2 degrees of freedom.

Solution.
Figure 6.4 compares the probability density functions for the normal and t distributions, the latter with 2 degrees of freedom.

```
x <- seq(-3, 3, length.out=100)
plot(x, dnorm(x), type="l")
lines(x, dt(x, df=2), lty=2)
```

A line for $p = 0.95$ can be useful

```
abline(v=qnorm(p=0.95), lty="dotted")
```

---

[5] http://climate.weather.gc.ca/glossary_e.html.

Fig. 6.4 Probability density functions for the normal distribution (solid line) and the t distribution with 2 degrees of freedom (dashed line)

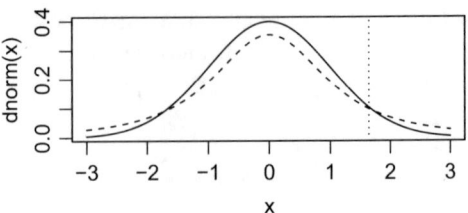

Exercise 2.31 on page 57. Write a function that computes measurement uncertainties assuming a t distribution.

Solution. The following is a function that does this, with `percent` being the desired coverage probability (Taylor and Kuyatt 1994); `na.omit()` is used to remove NA values.

```
uncertainty <- function(x, percent=95)
{
    x <- na.omit(x)
    k <- qt(p=1-(1-percent/100)/2, df=length(x)-1)
    u <- sd(x)
    k * u
}
```

(This function is used in Exercise 2.33.)

Exercise 2.32 on page 57. Write a function that plots error bars.

Solution. A simple solution is

```
errorBars <- function(x, ymin, ymax,dx=diff(range(x))/100, ...)
{
    segments(x, ymin, x, ymax, ...)
    segments(x-dx, ymin, x+dx, ymin, ...)
    segments(x-dx, ymax, x+dx, ymax, ...)
}
```

where dx has a default that might be useful, and ... contains graphical parameters (e.g. `col` or `lwd`) that are passed to `segments()`. (This function is used in Exercise 2.33.)

Exercise 2.33 on page 61. Show how `split()` and `laply()` can be used to produce a monthly climatology of a signal, and illustrate using the results of Exercises 2.31 and 2.32.

Solution. For illustration, a dataset named `rivsum.odf`, containing a century of monthly estimates of St Lawrence River discharge near Québec City (see, e.g., Bourgault and Koutitonsky 1999) was downloaded from a Department of Fisheries and Oceans website. This is in a specialized ODF format that is recognized by `read.oce()`, so

```
d <- read.oce("../data/rivsum.odf")
discharge <- d[["discharge"]]
time <- d[["time"]]
```

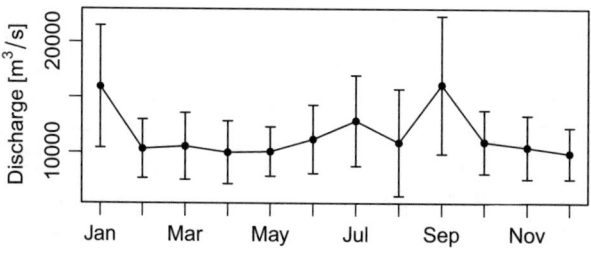

Fig. 6.5 Climatology of St Lawrence River discharge near Québec City (Exercise 2.33)

produces `discharge` in m³/s along with `time`, which is a `POSIXct` value from which the month may be extracted with `months()`, after which `split()` may be used to create a list that collects data into monthly categories. Analysing this list with `laply()` from the `plyr` package then yields monthly mean discarge and uncertainty, with

```
library(plyr)                              # for laply
dischargeSplit <- split(discharge, months(time))
y <- laply(dischargeSplit, mean)
unc <- laply(dischargeSplit, uncertainty)
ymin <- y - unc
ymax <- y + unc
```

where `uncertainty()` from Exercise 2.31 has been used. Incorporating the results of Exercise 2.32, Fig. 6.5 is made with

```
plot(1:12, y, pch=20, type="o", axes=FALSE,
    xlab="", ylab=expression("Discharge ["*m^3/s*"]"),
    ylim=range(c(ymin, ymax)))
errorBars(1:12, ymin, ymax)
box()
axis(2)
axis(1, 1:12, c("Jan", "Feb", "Mar", "Apr", "May", "Jun",
        "Jul", "Aug", "Sep", "Oct", "Nov", "Dec"))
```

where the axis is constructed manually to get month names instead of numbers. Note also the setting of `ylim` to encompass not just the data but also the error bars.

Exercise 2.34 on page 66. Contrast the residual plots produced by `plot()` for `linear` and `quadratic`.

Solution. With data and regressions as in the text, the solution is

```
plot(linear, which=1)
plot(quadratic, which=1)
```

with the resultant Fig. 6.6 suggesting that the quadratic model is preferable to the linear one, with lower residuals and a less pronounced and systematic pattern of deviations.

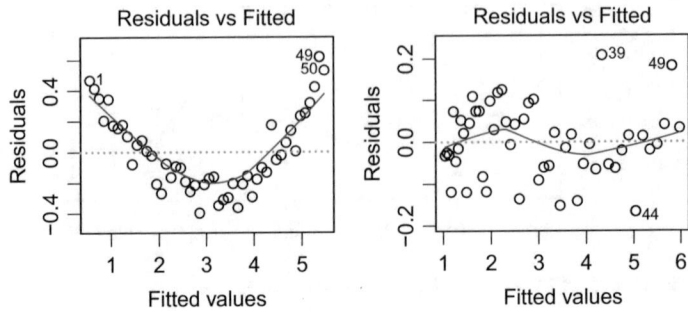

Fig. 6.6 Regression diagnostics for a linear and quadratic model (Exercise 2.34)

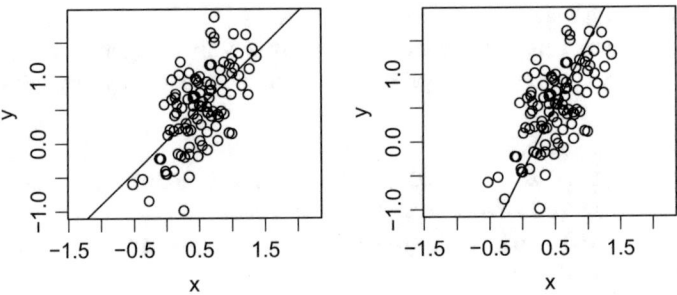

Fig. 6.7 Left: conventional linear fit; right: eigen-analysis fit (Exercise 2.35)

Exercise 2.35 on page 66. Use `eigen()` and `cov()` to draw a line that intersects the means of x and y, and that has the same slope as the principal eigenvector of the covariance matrix.

Solution. The first step is to construct artificial data

```
x <- seq(0, 1, length.out=100) + rnorm(100, sd=0.2)
y <- x + rnorm(100, sd=0.4)
```

and then plot a data cloud (Fig. 6.7)

```
par(mfrow=c(1, 2))
plot(x, y, asp=1)
abline(lm(y ~ x))
```

with aspect ratio equal to 1, assuming x and y to be of similar scale. (This assumption of similar scale, more particularly of errors in x and y, is central to this method.) The data are bound together into a data frame for use by `cov()`, and the desired eigenvector is stored in the first column of the value returned by `eigen()`. Thus the gist of the solution is

```
e <- eigen(cov(data.frame(x, y)))
```

The desired slope is the ratio of y and x components of the eigenvector

```
B <- e$vector[2,1] / e$vector[1,1]
```

Fig. 6.8 Geosecs station
235, data and nonlinear curve
fit (Exercise 2.36)

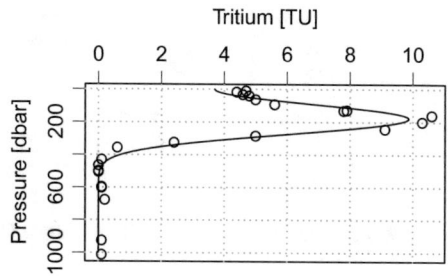

and so it only remains to solve for the intercept

```
A <- mean(y) - B * mean(x)
plot(x, y, asp=1)
abline(A, B)
```

Figure 6.7 suggests that the results of the eigen method pass through the data cloud better than those of conventional regression.

Exercise 2.36 on page 69. Extract tritium Tu and pressure p from the ocedata dataset geosecs235, and use nls() to fit the model

$$\text{Tu} = A \exp(-(p - p_0)^2/D^2) + A \exp(-(p + p_0)^2/D^2)$$

where A, p_0 and D are parameters to be inferred.

Solution. First, extract the data

```
data(geosecs235, package="ocedata")
Tu <- geosecs235[["tritium"]]
p <- geosecs235[["pressure"]]
```

and plot them (Fig. 6.8) to get an idea for starting values for parameters.

```
plotProfile(geosecs235, xtype="tritium", type="p")
```

The diagram suggests $A \sim 10$, $D \sim 100$ dbar and $p_0 \sim 200$ dbar, and

```
m <- nls(Tu~A*exp(-((p-p0)/D)^2)+A*exp(-((p+p0)/D)^2),
         start=list(A=10, p0=200, D=100))
pp <- seq(1000, 0, -10)
lines(predict(m, newdata=list(p=pp)), pp)
```

adds the corresponding solution to Fig. 6.8, which compares well with Figure 10 of Kelley and Van Scoy (1999).

Exercise 2.37 on page 71. Increase the value of n until the TukeyHSD diagram indicates that $T1$ and $T3$ are producing different values.

Solution. The code is the same as in the text, apart from changing n in the rnorm() call and removing the random-seed setting. Experimentation suggests that increasing n to 50 makes the difference between $T1$ and $T3$ visibly different from zero on the TukeyHSD() plot, and that values larger than 100 yield differences of twice the range of the parameter estimates. (This sort of analysis can be useful in planning experiments.)

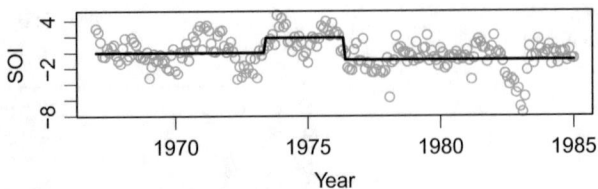

Fig. 6.9 Possible "regime shifts" in the Southern Oscillation Index (Exercise 2.38)

Exercise 2.38 on page 72. Use `ctree()` and `cpt.mean()` to examine the Southern Oscillation Index data (`soi` in the `oce` package) for regime shifts between 1967 and 1985.

Solution. The data may be subsetted to cover the suspected 1976–1977 shift discussed by Miller et al. (1994), and plotted in Fig. 6.9 with

```
data(soi, package="ocedata")
soi2 <- subset(soi, 1967 <= year & year <= 1985)
plot(soi2$year, soi2$index, col="darkgray",
     xlab="Year", ylab="SOI")
```

With default arguments, `ctree()` suggests shifts at years 1973.333 and 1976.333:

```
library(party)
regimes <- ctree(index~year, data=soi2)
lines(soi2$year, predict(regimes), lwd=3)
regimes@tree
1) year <= 1976.333; criterion = 1,
      statistic = 14.079
  2) year <= 1973.333; criterion = 0.999,
      statistic = 11.713
    3)* weights = 77
  2) year > 1973.333
    4)* weights = 36
1) year > 1976.333
  5)* weights = 104
```

but `cpt.mean()` in the `changepoint` package (Killick and Eckley 2014; Killick et al. 2016) suggests just one shift, at 1976.333

```
library(changepoint)
soi$year[cpt.mean(soi2$index)@cpts[1]]
[1] 1976.333
```

Using other data, Miller et al. (1994) identified a shift in 1976, i.e. in the year found by each of methods used here. However, there is a danger in over-interpreting results such as these; see, e.g., Cahill et al. (2015).

Exercise 2.39 on page 73. Use `integrate()` to calculate the perimeter of an ellipse of major axis $a = 2$ and minor axis $b = 1$.

Solution. The result is given by

$$4a \int_0^{\pi/4} \left(1 - \frac{a^2 - b^2}{b^2} \sin^2 \theta\right)^{1/2} d\theta \tag{6.3}$$

and this has numerical solution

```
a <- 2
b <- 1
integrate(function(t)
          4*a*sqrt(1-(a^2-b^2)/a^2*sin(t)^2),0, pi/2)
9.688448 with absolute error < 3.5e-10
```

which matches the solution from Sect. 2.3.11.1 to within 7×10^{-15}.

Exercise 2.40 on page 75. Use interp.surface to find water depth H under the mean Gulf Stream position as defined in the gs dataset of the ocedata package. Draw a map of the Gulf Stream location along with a graph of how H varies with distance along the path.

Solution. First, extract the mean position of the Gulf Stream path. The gs dataset holds longitude as a vector and latitude as a matrix (for various estimates of the position). The matrix can be averaged using lapply()

```
library(fields)
data(gs, package="ocedata")
lon <- rev(gs$longitude)
lat <- rev(apply(gs$latitude, 1, mean))
```

where rev() is used because the data are ordered from north to south. Next, draw the coastline in Lambert Conformal Conic projection (Fig. 6.10)

```
data(coastlineWorldMedium, package="ocedata")
par(mfrow=c(1,2), mar=c(3,3,1,1))
mapPlot(coastlineWorldMedium, proj="+proj=lcc
+lon_0=-65", grid=c(5, 5),
```

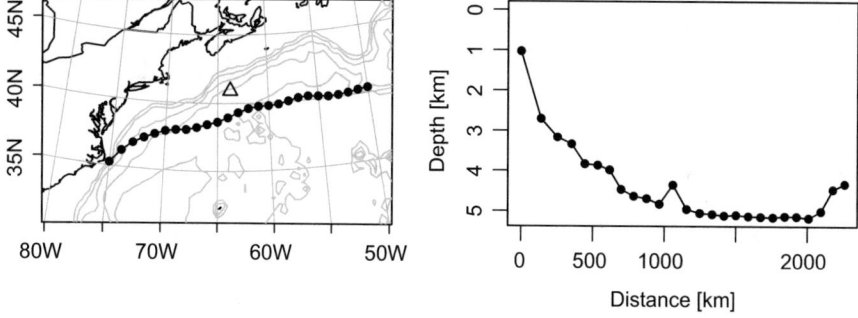

Fig. 6.10 Left: Gulf Stream position; right: water depth along trajectory (Exercise 2.40)

```
                longitudelim=c(280, 310),
                latitudelim=c(35, 43))
```
Then, superimpose depths in 1-km increments
```
data(topoWorld, package="oce")
z <- topoWorld[["z"]]
x <- topoWorld[["longitude"]]
y <- topoWorld[["latitude"]]
mapContour(x, y, z, levels=seq(-6000,-1000,1000),
col="gray")
```
and add the path of the Gulf Stream.
```
mapLines(lon, lat, type="o", pch=20)
```
As a matter of historical interest, Milne (1867) estimated the northern limit of the Gulf Stream between Halifax and Bermuda as 40°56'N at 63°45'W
```
mapPoints(-(63+45/60), 40+56/60, pch=2)
```
The final step is to draw the graph of interpolated depth under the Gulf Stream, a task made easy with interp.surface()
```
H <- -0.001*interp.surface(list(x=x,y=y,z=z),
cbind(lon,lat))
distance <- geodDist(lon, lat, lon[1], lat[1])
plot(distance, H, xlab="Distance [km]",
        ylab="Depth [km]", ylim=c(max(H), 0), type="o",
        k pch=20)
```

Exercise 2.41 on page 77. Contrast the predictions of interpolating and smoothing splines for the turbulence data.
Solution.
A comparison is shown in Fig. 6.11, using data
```
data(turbulence, package="ocedata")
```
from which components are extracted with
```
k <- turbulence$k
phi <- turbulence$phi
y <- k^2 * phi
```

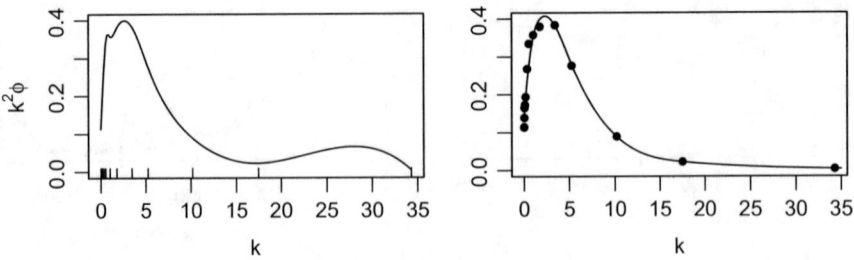

Fig. 6.11 Interpolating and smoothing splines with turbulence data (Exercise 2.41)

A simple plot will have data point obscuring the spline, so use `type="n"` to get axes without the data points, and use `rug()` to place ticks inside the bottom of the plot box.

```
par(mfrow=c(1,2))
plot(k, y, ylab=expression(k^2*phi), ylim=c(0,0.40),
   type="n") rug(k, side=1, ticksize=0.06, lwd=1)
```

Next, add an interpolating spline with $n = 200$ segments

```
n <- 200
lines(spline(k, y, n=n))
```

A sharp eye will see an aphysical wiggle at $k \approx 1$, and even a glance reveals a much larger one in the range $17 < k < 35$. By contrast, a smoothing spline

```
plot(k, y, ylab="", ylim=c(0,0.40), pch=20)
lines(predict(smooth.spline(k,y,df=7),seq(0,35,length.
   out=n)))
```

yields a curve with a smoother character that may make more sense. The cost of using a smoothing spline is a misfit with the data, and an occasional need to alter default smoothing parameters, e.g. the setting of `df` here. In many applications, smoothing splines produce results that match physical expectations, e.g. the avoidance of wiggles is helpful in the calculation of the square of buoyancy frequency by `swN2()`.

Exercise 2.42 on page 77. Create a function returning the prediction of a smoothing spline, and use it to calculate ϵ as in Sect. 2.6.4.
Solution. First, construct the spline function

```
s <- smooth.spline(k, k^2 * phi)
```

and then construct a wrapper around its `predict()` function

```
f <- function(x) predict(s, x)$y
```

at which point it makes sense to continue the application as in the main text

```
15 * nu * integrate(f, min(k), max(k))$value
 [1] 0.64018
```

This value is within 4.9% of that reported by Grant et al. (1962), an improvement over the estimate using piecewise-linear interpolation, in Sect. 2.6.4.

Exercise 2.43 on page 81. Read the Dalhousie-WHOI route using the XML package.
Solution. First, parse the file

```
library(XML)
p <- xmlParse("../data/dalwhoi.kml")
```

and extract the root element

```
r <- xmlRoot(p)
```

The next step is to discover what `r` contains. This can be done in R, but readers might find it faster to use an XML application or to examine the data file in a text editor. In any case, the following yields longitude and latitude along the route (note that it's also possible to name indices to `r`).

Table 6.1 Sampling of a cnv-format CTD file

Line	Contents
1	`* Sea-Bird SBE 25 Data File:`
2	`* FileName = C:\SEASOFT3\BASIN\BED0302.HEX`
11	`** Latitude: N44 41.056`
12	`** Longitude: w63 38.633`
19	`# name 2 = pr: pressure [db]`
42	`*END*`
43	`130 129.000 1.480 1.468 14.2245 29.9210 0.000e+00`

```
a <- xmlValue(r[["Document"]][[7]][[4]][[2]], trim=TRUE)
b <- strsplit(a, "\n")[[1]]
library(plyr)
loc <- laply(b, function(x) as.numeric(strsplit(x,",")[[1]]))
names(loc) <- c("lon", "lat", "unknown")
```

Exercise 2.44 on page 82. Read the sample CTD file `ctd.cnv`, skipping the header and naming the columns.

Solution. Loading the file

```
file <- system.file("extdata", "ctd.cnv",
package="oce")
lines <- readLines(file)
```

yields results as sampled in Table 6.1. Examination suggests that the line containing "`*END*`" flags the end of the header, so

```
end <- grep("*END*", lines)
data <- read.table(text=tail(lines, -end))
```

will read the columnar data. The next step is to find the data names

```
nameLines <- lines[grep("name", lines)]
names <- gsub("^# name .* = (.*):.*$", "\\1",
nameLines)
names
[1] "scan"  "timeS"  "pr"    "depS"  "t068"  "sal00"
"flag"
```

and to rename them, perhaps with, e.g.

```
names[names=="pr"] <- "pressure"
```

or *en masse*, with e.g.

```
names(data) <- c("scan", "time", "pressure", "depth",
                 "temperature", "salinity", "flag")
```

Readers with CTD experience will know that data headers vary widely, and may appreciate the fact that `oce` automatically recognizes well over 100 hydrographic variable names (and associates them with data-quality flags, if these are present in the `.cnv` file).

Exercise 2.45 on page 82. Read the SOI data using `scan()`.

Solution. The first step is to download the data from the source listed on page 80. Time and missing-value can be identified simply, and time constructed with `seq()`

```
d <- scan("../data/soi.dat")
n <- length(d) / 13 - 1
year <- 1/24 + seq(from=d[1], to=d[1] + n+11/12,
by=1/12)
SOI <- d[d < 1800]
SOI[SOI < -90] <- NA
```

Exercise 2.46 on page 85. From https://rbr-global.com/support/matlab-tools, get `RSKtools` and extract time and temperature from the SQLite file named `sample.rsk`.

Solution. These data are not stored in order of increasing time, so it is necessary to order by time. This could be done within R, but it is just as easy to do it in the SQLite query. The first step is to load the file

```
library(RSQLite)
f <- dbConnect(SQLite(), "../data/sample.rsk")
```

and then extract the time in milliseconds

```
d <- dbGetQuery(f,
        "select tstamp from data order by tstamp")
t <- numberAsPOSIXct(d[[1]] / 1000)
```

The table named `channels` tells what is stored in the file, indicating that temperature may be retrieved with

```
T <- dbGetQuery(f,
        "select channel02 from data order by tstamp")
        [[1]]
```

Exercise 2.47 on page 86. Plot SST contours with a coastline.

Solution. Constructing Fig. 6.12 is best done by first drawing the coastline, because oce provides a generic function for coastline objects that will set up sensible axes (e.g. limiting latitude range). So, the first plotting step is

```
data(coastlineWorld, package="oce")
plot(coastlineWorld)
```

A 5-degree version of the 2013 World Ocean Atlas may be read with

```
library(ncdf4)
con <- nc_open("../data/woa13_decav_t00_5dv2.nc")
lon <- ncvar_get(con, "lon")
lat <- ncvar_get(con, "lat")
SST <- ncvar_get(con, "t_mn")[,,1]
```

but this has longitude ranging from 0 to 360, so conversion to the −180 to 180 convention used for the coastline is required, with

```
lon <- ifelse(lon > 180, lon - 360, lon)

contour(lon, lat, SST, add=TRUE, labcex=1)
```

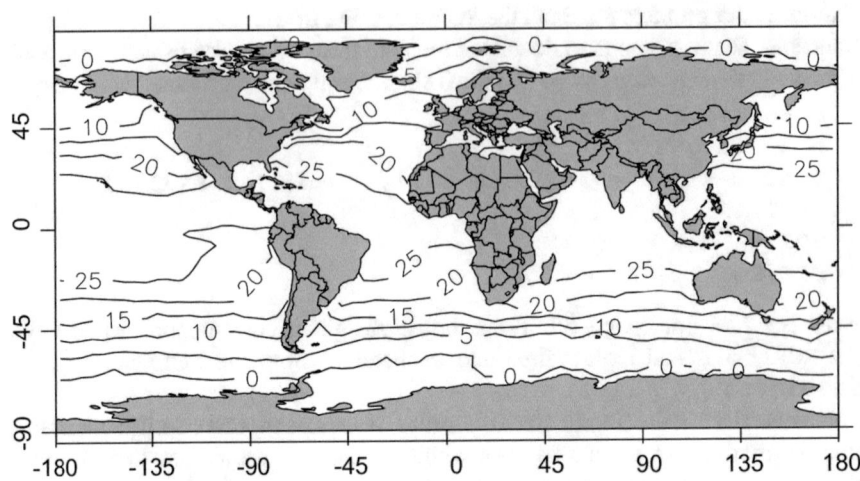

Fig. 6.12 Contours of sea surface temperature from the World Ocean Atlas

Exercise 2.48 on page 89. Extend the CTD-edit `shiny` app to read data from a file, set flags for bad data and save the result to a file.

Solution. Adding a call to `fileInput()` in `ui.R` will let the user either type a file name or use a system browser to identify a file. The `datapath` item in the returned value is the name of the file to be used. The contents of that file should be used to define "`ctd`," and checks will need to be set up to prevent using that quantity before it is defined. The flagging and saving steps can be accomplished together. Adding

```
actionButton("save_file", "Save Output")
```

in `ui.R` sets up a linkage to a function to save the output, and defining that function with

```
observeEvent(input$save_file, {
  ctd[["salinityFlag"]] <- ifelse(vals$keep, 2, 4)
  save(ctd, file="ctd_edited.rda")
})
```

in `server.R` accomplishes the task using World Hydrographic Program error codes of 2 for good data and 4 for bad data. Here, a simple filename is used, but in practice it would make sense to use a name derived from the input filename. The object could be trimmed of bad data with `handleFlags()`.

6.2   Exercises in Chap. 3

Exercise 3.1 on page 97. Use the generic `plot()` for CTD objects, to produce a version of Fig. 3.1 using the UNESCO equation of state instead of the default TEOS-10 version.
Solution. Use `plot(ctd, eos="unesco")`

Exercise 3.2 on page 97. (a) Calculate the density of seawater at pressure 100 dbar, salinity 34 PSU, and temperature 10°C. (b) What temperature would the parcel have if raised adiabatically to the surface? (c) What density would it have if raised adiabatically to the surface? (d) What density would it have if lowered about 100 m, increasing the pressure to 200 dbar? (e) Draw a blank T-S diagram with S from 30 to 40 PSU and T from -2 to 20°C.
Solution. The `oce` package provides the functions required for this exercise, all starting with the characters "sw" in their names. Since longitude and latitude are not given, the UNESCO equation of state will be used. This can be specified in a startup file, but this illustration uses the `eos` argument to the various `sw` functions.

(a) Density $\rho = \rho(S, T, p)$ is
```
swRho(34, 10, 100, eos="unesco")
[1] 1026.624
```
(b) Potential temperature $\theta = \theta(S, T, p)$ is
```
swTheta(34, 10, 100, eos="unesco")
[1] 9.988599
```
(c) Use θ for temperature
```
swRho(34, swTheta(34, 10, 100, eos="unesco"), 0,
eos="unesco")
[1] 1026.173
```
(d) Use θ, with reference pressure of 200 dbar, for temperature
```
theta <- swTheta(34, 10, 100, 200, eos="unesco")
swRho(34, theta, 200, eos="unesco")
[1] 1027.074
```
(e) Use white for the dots
```
plotTS(as.ctd(c(30,40),c(-2,20),rep(0,2)),
eos="unesco", type="n")
```

Exercise 3.3 on page 98. Use `propagate` from the `propagate` package to estimate typical CTD salinity uncertainty.
Solution. The SeaBird 911plus specifications state uncertainties of 0.0003 S/m for conductivity, 0.001 °C for temperature and 0.015% percent of range for pressure. Dividing the conductivity by 4.29140 S/m to get a conductivity ratio (Culkin and Smith 1980) and assuming a 1400dbar-scale pressure sensor,
```
library(propagate)
Ce <- 0.0003 / 4.29140
Te <- 0.001
```

```
pe <- 0.015e-2*1400
d <- data.frame(C=c(1,Ce),T=T90fromT68(c(15,Te)),
p=c(0,pe))
propagate(expression(swSCTp(C,T,p,eos="unesco")),
data=d)
```

The resultant Monte Carlo 95% confidence interval is 34.994 to 35.006, suggesting a CTD salinity uncertainty of ±0.005.

Exercise 3.4 on page 100. Map ocean-surface density.
Solution. The procedure (results not reproduced here) is the same as for Fig. 3.3, although first density must be computed. Note also the use of quantile() to prevent brackish waters from setting the scale.

```
data(coastlineWorld, package="oce")
data(levitus, package="ocedata")
ssrho <- swRho(levitus$SSS, levitus$SST, 0,
eos="unesco")
cm <- colormap(quantile(ssrho, c(0.025, 0.975),
              na.rm=TRUE), col=oce.colorsJet)
drawPalette(colormap=cm)
mapPlot(coastlineWorld, projection="+proj=moll")
mapImage(levitus$longitude, levitus$latitude, ssrho,
        colormap=cm)
mapGrid()
mapLines(coastlineWorld)
```

Exercise 3.5 on page 100. Use mapPlot() to draw a world coastline with the Robinson projection, and trace the 1700s H.M.S. Endeavour cruise.
Solution. Appendix C reveals that the codename of this projection is robin, so the map (Fig. 6.1) is drawn with

```
data(coastlineWorld, package="oce")
mapPlot(coastlineWorld, projection="+proj=robin",
        drawBox=FALSE)
data(endeavour, package="ocedata")
mapPoints(endeavour$longitude, endeavour$latitude,
          pch=20, cex=0.5)
```

This projection was used by the National Geographic Society for world maps, prior to a switch made to the Winkel Tripel near the end of the twentieth century. Comparison of Figs. 6.1 and C.1 reveals that the two projections differ mainly at high latitudes.

6.3   Exercises in Chap. 4

Exercise 4.1 on page 108. Use `loess()` to fit locally weighted polynomial models
(Sect. 2.6.6) of PO_4 as a function of NO_3, and vice versa.
Solution. The first step is to recreate Fig. 4.2, as Fig. 6.13, using the same steps as
in Sect. 4.1, but with `pch=20` in the `plot()` call, for smaller symbols. Next, add
a `loess` fit for NO_3 as a function of PO_4

```
l1 <- loess(NO3 ~ PO4, data=redfieldNP)
lines(redfieldNP$PO4, predict(l1))
```

and the opposite, using `order()` to put the data into order

```
l2 <- loess(PO4 ~ NO3, data=redfieldNP)
o <- order(redfieldNP$NO3)
lines(predict(l2)[o], redfieldNP$NO3[o],
lty="longdash")
```

Exercise 4.2 on page 108. Alter the `lm()` call for the Redfield ratio fit, to test
whether the slope might be 20.
Solution. The *p* value for the slope, labelled "`Pr`" in the summary, is in comparison
to zero slope. To compare with a slope of 20, one may create a transformed variable,
either directly

```
NO3new <- redfieldNP$NO3 - 20 * redfieldNP$PO4
m <- lm(NO3new ~ PO4 - 1, data=redfieldNP)
```

or indirectly (skipping the creation of a variable not needed later)

```
m <- lm(NO3 - 20*PO4 ~ PO4 - 1, data=redfieldNP)
```

In either case, `summary(m)` reveals output including the following lines

```
      Estimate Std. Error t value Pr(>|t|)
PO4   -0.8440      0.2726  -3.097  0.00245 **
```

suggesting a slope other than 20.

Fig. 6.13 `loess` predictions
for NO_3 as a function of PO_4
(solid) and the reverse
(dashed)

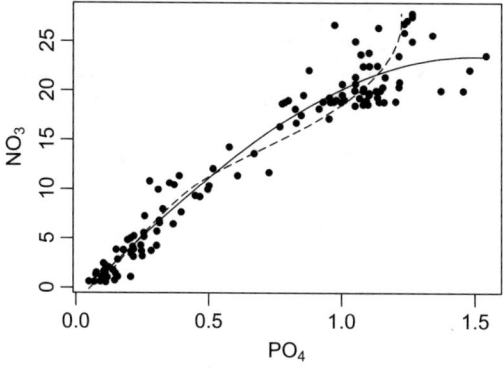

Exercise 4.3 on page 108. Calculate the slope in Fig. 4.2 using ridge, robust and resistant regression, using functions `lm.ridge()`, `rlm()` and `lqs()` from the MASS package, respectively.

Solution. With data loaded as in the text, the slopes for these methods are as follows: the predictions lie between the lines in Fig. 4.2.

```
library(MASS)
as.numeric(coef(lm.ridge(NO3 ~ PO4 - 1,
data=redfieldNP)))
[1] 19.15596

as.numeric(coef(rlm(NO3 ~ PO4 - 1, data=redfieldNP)))
[1] 19.22685

as.numeric(coef(lqs(NO3 ~ PO4 - 1, data=redfieldNP)))
[1] 19.72996
```

Exercise 4.4 on page 111. Use `lsoda()` to solve the NPZ equations as expressed in Chapter 4 of Sarmiento and Gruber (2006)

$$\frac{dN}{dt} = P\left(-V_{max}\frac{N}{K_N+N} + \mu_P\lambda_P\right) + Z\mu_Z\left[(1-\gamma_Z)g\frac{P}{K_P} + \lambda_Z\right]$$

$$\frac{dP}{dt} = P\left(V_{max}\frac{N}{K_N+N} - \lambda_P - \frac{gZ}{K_P}\right)$$

$$\frac{dZ}{dt} = Z\left(\gamma_Z g\frac{P}{K_P} - \lambda_Z\right)$$

(6.4)

with $V_{max} = 1.4\,\mathrm{d}^{-1}$, $K_N = 0.1\,\mathrm{mmol/m^3}$, $\mu_P = 1$, $\lambda_P = 0.05\,\mathrm{d}^{-1}$, $\mu_Z = 1$, $\gamma_Z = 0.4$, $g = 1.4\,\mathrm{d}^{-1}$, $K_P = 2.8\,\mathrm{mmol/m^3}$, and $\lambda_Z = 0.12\,\mathrm{d}^{-1}$. Use initial conditions $N = 10$, $P = 3$ and $Z = 2$, and plot the results over a month.

Solution. The first step is to define the differential equations

```
library(deSolve)
NPZ <- function(t, y, parms, ...) {
    N <- y[1] ; P <- y[2] ; Z <- y[3] # rename for clarity
    N0 <- N/(parms$KN+N)
    dNdt<-P*(-parms$Vmax*N0+parms$muP*parms$lambdaP) +
          parms$muZ*Z*((1-parms$gammaZ)*parms$g*P/parms$KP+
                parms$lambdaZ)
    dPdt<-P*(parms$Vmax*N0-parms$lambdaP -
            (parms$g/parms$KP)*Z)
    dZdt<-Z*(parms$gammaZ*parms$g*P/parms$KP-parms$lambdaZ)
    list(c(dNdt=dNdt, dPdt=dPdt, dZdt=dZdt))
}
```

and parameters

```
parms <- list(Vmax=1.4, KN=0.1, muP=1, lambdaP=0.05,
              muZ=1,
              gammaZ=0.4, g=1.4, KP=2.8, lambdaZ=0.12)
```

Next, set initial conditions and solve

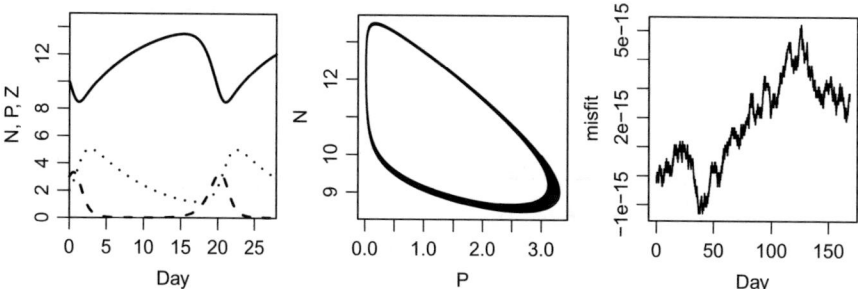

Fig. 6.14 NPZ simulation results (Exercise 4.4). Left: variation of N (solid), P (long dash) and Z (short dash) over a month. Middle: covariation of N and P over 6 months, showing slow variation of cycling. Right: fractional error in $N + P + Z$ over 6 months

```
IC <- c(N=10, P=3, Z=2)
times <- seq(0, 28, 1/24)
solution <- lsoda(IC, times, func=NPZ, parms=parms)
```
The solution is plotted with
```
plot(solution[,1], solution[,2], xaxs="i", yaxs="i",
     ylim=c(0, sum(IC)), xlab="Day", ylab="N, P, Z",
     type="l", lwd=2, lty=1)
lines(solution[,1], solution[,3], lwd=2, lty=2)
lines(solution[,1], solution[,4], lwd=2, lty=3)
```
and the resultant left panel of Fig. 6.14 shows a reduction in nutrient for about a day, with a phytoplankton bloom, followed by an increase in zooplankton concentration and then a decline of both phytoplankton and zooplankton. The pattern appears to repeat, but a check in state space over 6 months (middle panel) suggests a slow variation in the cycle[6]

One way to test for numerical errors is to compute $N + P + Z$, which should be constant in (6.4) since $\mu_Z = \mu_P = 1$. A six-month simulation
```
misfit <- (sum(IC) - apply(solution[,-1], 1, sum))/
sum(IC) plot(solution[,1], misfit,xlab="Day",type="l")
```
(right panel of Fig. 6.14) suggests that lsoda() conserves $N + P + Z$ to under 1 part in 10^{14} on a 64-bit computer, a compelling reason to use this function instead of a crude Euler-step integrator.

Exercise 4.5 on page 111. Use lsoda() to develop a numerical solution to the wave equation $d^2\eta/dt^2 + \omega^2\eta = 0$ with frequency $\omega = 1\,\text{s}^{-1}$ during $0 \le t \le 2\pi$, with initial condition $\eta = 0$ and $d\eta/dt = 1$ at $t = 0$.
Solution. Introducing ζ as a new variable, we get a pair of first-order equations

---

[6]Such simulations are useful for exploring the nature of the equations, but realistic applications should include changing background conditions and forcing.

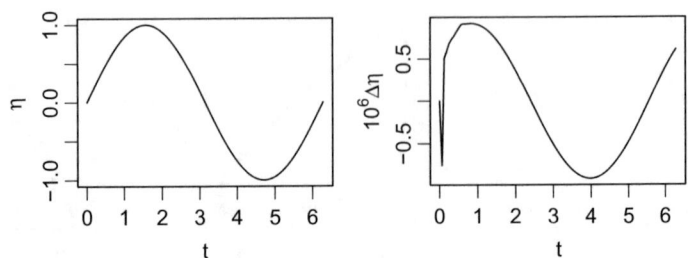

Fig. 6.15 Left: numerical solution to wave equation; right: 10^6 times error. (Exercise 4.5)

$$\frac{d\eta}{dt} = \zeta \quad \text{and} \quad \frac{d\zeta}{dt} = -\omega^2\eta \qquad (6.5)$$

which may be expressed in R as

```
DE <- function(t, y, parameters)
      list(c(y[2], -parameters$omega*y[1]))
```

where y[1] stands for η and y[2] stands for ζ. The solution is simpler than in the previous exercise:

```
library(deSolve)
IC <- c(0, 1)
t <- seq(0,2*pi,length.out=100)
parameters <- list(omega=1)
soln <- lsoda(IC, t, DE, parameters)
```

Since soln holds t, η and ζ, $\eta(t)$ and its deviation from the analytical solution $\sin(t)$ are plotted in Fig. 6.15 with

```
par(mfrow=c(1, 2))
plot(soln[,1], soln[,2],
     xlab="t", ylab=expression(eta), type="l")
plot(soln[,1], 1e6*(soln[,2]-sin(soln[,1])),
     xlab="t", ylab=expression(10^6*Delta*eta),
     type="l")
```

Exercise 4.6 on page 113. Explore type-II regression with the lmodel2 package. **Solution.** The documentation for lmodel2() should prove useful to readers who are unfamiliar with type-II regression. Figure 6.16 starts by mimicking Fig. 4.4, drawing the data and Wilson's line. Regression through the origin is not offered by lmodel2, so the formula used to describe the regression is different from that in Sect. 4.3. The arguments range.x and range.y are used by the "RMA" (ranged major axis) method, since both distance and age have meaningful zero values in this case.

Fig. 6.16 Re-analysis of seafloor-spreading data from Table 1 of Wilson (1963), using various alternative regression methods; compare with Fig. 4.4

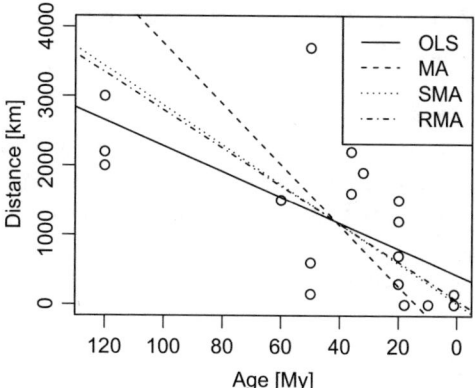

```
library(lmodel2)
m <- lmodel2(Distance ~ Age, data=wilson,
             range.y="relative", range.x="relative")
```

The value returned by lmodel2() includes regression.results, a matrix summarizing the coefficients along with other things. To draw the results, only three columns from the matrix are needed.

```
rr <- m$regression.results
n <- dim(rr)[1]
for (method in 1:n)
    abline(rr[method,2], rr[method,3], lty=method)
legend("topright", lty=1:n, legend=rr[,1])
```

Readers might find it informative to compare the results designated SMA with those found with sma() in the smatr package, as presented in Sect. 4.3.

Exercise 4.7 on page 116. Use try() to skip past errors in nls(), so that a bootstrap estimate of H can be done.

Solution. The key is to wrap the function call in a try block. The z and T vectors defined in the text may be bound into a matrix with

```
zT <- cbind(z, T)
```

that is to be provided to a fitting function

```
safeFit <- function(zT, i)
{
    try({
        m <- nls(zT[i,2]~Td+(Ts-Td)*exp(zT[i,1]/h),
                 start=list(Ts=5, Td=1, h=1000))
        coef(m)[["h"]]
    },
    silent=TRUE)
}
```

With these definitions, load `boot` and perform the calculation with 100 replicates (output not reproduced here).

```
library(boot)
set.seed(47)
b <- boot(data=zT, statistic=safeFit, R=100)
```

after which a confidence interval is constructed with

```
boot.ci(b, type="basic") # results not printed
```

which yields a 95% CI of $747 < h < 1029$ m. (Note that this range encloses the Munk (1966) value of 0.77 km.)

Exercise 4.8 on page 116. Specify gradients to `nls()` to fit the `munk` data.
Solution. The gradient of the model Eq. (4.4) has components

$$\frac{\partial \theta}{\partial \theta_S} = 1 - \exp(z/h)$$

$$\frac{\partial \theta}{\partial \theta_D} = \exp(z/h) \qquad\qquad (6.6)$$

$$\frac{\partial \theta}{\partial h} = -\frac{z}{h^2}(\theta_D - \theta_D)\exp(z/h)$$

which can be provided to `nls()` with

```
munk <- function(thetaS, thetaD, h)
{
    E <- exp(z / h)
    prediction <- thetaD + (thetaS - thetaD) * E
    gthetaS <- E
    gthetaD <- 1 - E
    gh <- -z / h^2 * (thetaS - thetaD) * E
    gradient <- cbind(gthetaS, gthetaD, gh)
    attr(prediction, "gradient") <- gradient
    prediction
}
m <- nls(theta~munk(thetaS, thetaD, h),
         start=list(thetaS=5, thetaD=1, h=1000))
```

and a check with

```
coef(m)
    thetaS        thetaD              h
 11.231380     1.165834  844.653897
```

indicates that the results match those listed on page 115.

6.4 Exercises in Chap. 5

Exercise 5.1 on page 122. Use sub() to create a series of PDF files with names that map to the names of data files.
Solution.
```
for (file in list.files(path=".", pattern=".cnv$")) {
    pdf(gsub(".cnv$", ".pdf", file))
    plot(ctdTrim(read.ctd(file)))
    dev.off()
}
```

Exercise 5.2 on page 124. Explore the sensitivity of buoyancy frequency, calculated with swN2(), to the argument df.
Solution. A graphical approach can be helpful, and Fig. 6.17 starts with a default N^2 profile, computed from trimmed and decimated data, starting with
```
data(ctdRaw, package="oce")
ctd2 <- ctdTrim(ctdRaw, "range", parameters=list
               (item="scan", from=130, to=320))
ctd3 <- ctdDecimate(ctd2, p=seq(0,50,1), "boxcar")
plot(ctd3, which="N2", col.N2="black")
```
after which an N^2 profile with specified df may be added simply
```
lines(swN2(ctd3, df=10), ctd3[["pressure"]],
      lty="dashed")
```
In this particular case, setting df=10 yields a more variable N^2 profile, because the default swN2() call made by plot has computed a default df value that is less than 10. In some instances, there will be extra information guiding a choice of df (or similar parameters used by smooth.spline(), or other smoothing functions chosen by the user). For example, some analysts might examine the artificial density profile computed by integrating N^2 over depth, to see if the calculation has captured

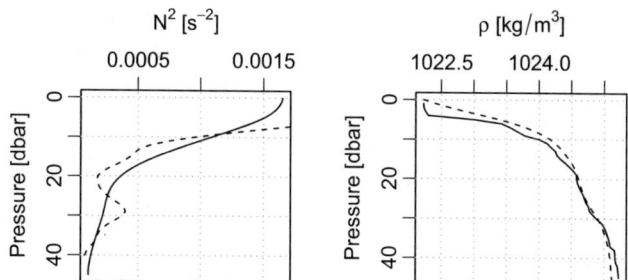

Fig. 6.17 Left panel: profile of buoyancy frequency N^2 computed by default with plot() (solid), and computed with df set to 10 (dashed). Right panel: density profile (solid), along with profile computed by integrating the N^2 profile calculated with df=10

features of interest, e.g. pycnocline location in some cases, mixed-layer depth in other. A simple procedure for estimating such an artificial profile follows.

```
plot(ctd3, which="density")
dz <- mean(diff(swZ(ctd3)), na.rm=TRUE)
rho0 <- mean(swRho(ctd3), na.rm=TRUE)
rhom <- min(swRho(ctd3), na.rm=TRUE)
g <- gravity() # may specify latitude here
N2 <- swN2(ctd3, df=10)
N2[is.na(N2)] <- 0 # just in case
rho <- rhom - dz * (cumsum(N2) - N2[1]) * rho0 / g
lines(rho, ctd3[["pressure"]], lty="dashed")
```

Exercise 5.3 on page 124. Plot salinity and temperature profiles for the ctd dataset within 3 dbar of the pycnocline centre.

Solution. First, load the data and compute N^2 values

```
data(ctd, package="oce")
N2 <- swN2(ctd)
```

The pycnocline centre might be estimated from the pressure at which N^2 is maximum

```
p0 <- ctd[["pressure"]][which.max(N2)]
```

after which focus is directed to the suggested region with

```
pycnocline<-subset(ctd, p0-2 <= pressure & pressure
<= p0+2)
```

and then Fig. 6.18 is completed with

```
par(mfrow=c(1, 2))
plot(pycnocline, which="salinity", type="o",
yaxs="i") abline(h=p0, lty=2)
plot(pycnocline, which="temperature", type="o",
yaxs="i") abline(h=p0, lty=2)
mtext(sprintf("%.1f dbar", p0), side=4, at=p0)
```

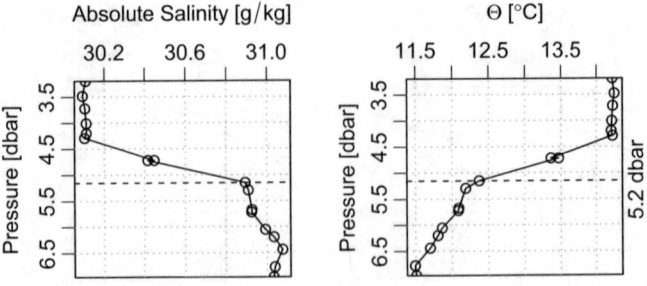

Fig. 6.18 Absolute Salinity and Conservative Temperature profiles, near the pycnocline of the ctd dataset (Exercise 5.3)

Exercise 5.4 on page 124. Use `ctdTrim()` and `plotScan()` together, to trim `ctdRaw` to just the downcast portion.

Solution.

```
data(ctdRaw, package="oce")
plotScan(ctdRaw)
```

where the value of `which` indicates that a graph of pressure versus index is to be shown. The next step is to bracket the downcast in terms of index, and then to narrow this bracket based on observation of the graph.

Some users will find it convenient to determine the bracket by a visual inspection of the graph, while others may prefer to use `locator()` to select values by clicking on the graph, and still others will prefer to narrow in on the index by drawing on the graph, e.g. `abline(v=100)` shows that the start of the downcast is definitely after index 100. The last two approaches can be combined with, e.g.,

```
x <- round(locator(1)$x); abline(v=x); mtext(x, at=x,
             side=3)
```

where the semicolons make it easy to repeat the three steps in an interactive session, simply by striking the up-arrow key.

At this stage, it is convenient to start narrowing the focus. The use of `plotScan()`, which puts scan number on the x axis, is handy for this task. Using the arrow keys in an R console window provides an easy way to find appropriate limits on scan number, in a few keystrokes, e.g. the sequence

```
plotScan(ctdTrim(ctdRaw, "range",
               parameters=list(item="scan",from=100,
               to=400)))
plotScan(ctdTrim(ctdRaw, "range",
               parameters=list(item="scan",from=150,
               to=400)))
```

suggests a starting `scan` value of perhaps 130, and a similar sequence suggests an ending scan of 300. This second value is open to argument; the author settled upon it in order to achieve a roughly linear increase of pressure within the range of scan numbers. How such decisions should be made depends upon the instrumentation used, e.g. differing for pumped and unpumped conductivity sensors.

A wider data window is used by `ctdTrim()`, as readers can verify with

```
plotScan(ctdTrim(ctdRaw, method="sbe"))
```

Exercise 5.5 on page 128. Use the Chu and Fan (2010b) method on the `ctd` dataset.

Solution. A solution may be written as a function that returns not just the mixed layer depth, but also some other interesting elements of the calculation. The argument n determines the number of data levels to examine below the focus depth, while `variable` names the hydrographic variable to be examined, with temperature, as the default, `"sigmaTheta"` for σ_θ, etc.

```
MLDchu <- function(ctd, n=5, variable="temperature")
{
    pressure <- ctd[["pressure"]]
```

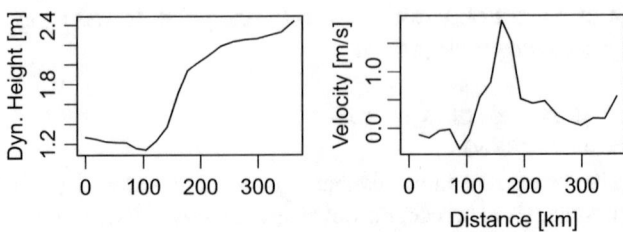

Fig. 6.19 Dynamic height and geostrophic velocity of the Gulf Stream (Exercise 5.6)

```
x <- ctd[[variable]]
ndata <- length(pressure)
E1 <- rep(NA, ndata)
E2 <- rep(NA, ndata)
E2overE1 <- rep(NA, ndata)
kstart <- min(n, 3)
for (k in seq(kstart, ndata - n, 1)) {
    above <- seq.int(1, k)
    below <- seq.int(k + 1, k + n)
    fit <- lm(x ~ pressure, subset=above)
    E1[k] <- sd(predict(fit) - x[above])
    pBelow <- data.frame(pressure=pressure[below])
    E2[k] <-abs(mean(predict(fit,newdata=pBelow)-x[below]))
    E2overE1[k] <- E2[k] / E1[k]
}
MLDindex <- which.max(E2overE1)
return(list(MLD=pressure[MLDindex], MLDindex=MLDindex,
            E1=E1, E2=E2, E2overE1=E2overE1))
}
```

Using the function to measure MLD is simple, e.g. the following plots the bottom-right panel of Fig. 5.4 on page 127.

```
data(ctd, package="oce")
mld <- MLDchu(ctd)
plotProfile(ctd, xtype="temperature", ylim=c(15, 0))
abline(h=mld$MLD, lwd=2, lty="dashed")
```

Exercise 5.6 on page 129. Plot dynamic height and geostrophic velocity across the Gulf Stream.

Solution. Start by subsetting and reordering the stations.

```
data(section, package="oce")
GS <- subset(section, -73.2<longitude &
                longitude<(-69.4))
GS <- sectionSort(GS, by="longitude")
```

Calculate dynamic height and plot it in Fig. 6.19.

```
par(mfrow=c(1,2), mar=c(3,3,0.5,0.5), mgp=mgp)
dh <- swDynamicHeight(GS)
```

Fig. 6.20 Split-depth temperature section of the Gulf Stream (Exercise 5.7)

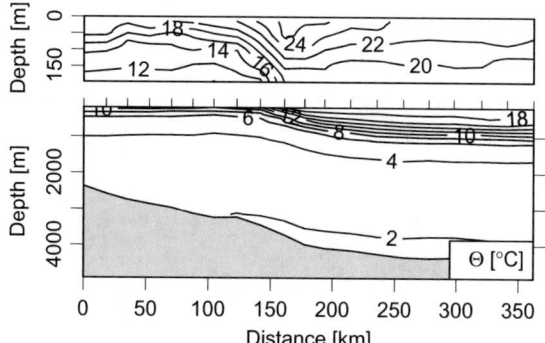

```
plot(dh$distance, dh$height, type="l",
     xlab="", ylab="Dyn. Height [m]")
```
then calculate and plot the geostrophic velocity
```
lat <- GS[["station", 1]][["latitude"]]
f <- coriolis(lat)
g <- gravity(lat)
v <- diff(dh$height)/diff(dh$distance) * g / f / 1e3
v <- c(NA, v) # to get same length as distance
plot(dh$distance, v, type="l",
     xlab="Distance [km]", ylab="Velocity [m/s]")
```

Exercise 5.7 on page 129. Plot a split-depth temperature section for the Gulf Stream, with a panel for variation in the top 200 m and another for variation below. Use `layout()` to make panels of unequal height.
Solution. Continuing with GS, Fig. 6.20 is constructed with
```
top <- sectionGrid(GS, p=seq(0, 200, 5))
bottom <- sectionGrid(GS, p=seq(200, 5000, 100))
layout(matrix(1:2, nrow=2), widths=1,
heights=c(0.25, 0.75))
plot(top, which="temperature", mar=c(0, 3, 1, 1),
     axes=FALSE, legend.loc="")
axis(2, at=pretty(par("usr")[3:4]))
plot(bottom, which="temperature",mar=c(3, 3, 1.0, 1))
```
where `axis()` has been provided with labels, since the default in this small domain would otherwise yield just one labelled depth.

Exercise 5.8 on page 134. Create a `coplot` of the `section` dataset, showing T–S dependence as a function of latitude and longitude.
Solution. Continuing with the `section` dataset,
```
data(section, package="oce")
sec <- handleFlags(section)          # remove bad data
S <- sec[["salinity"]]
```

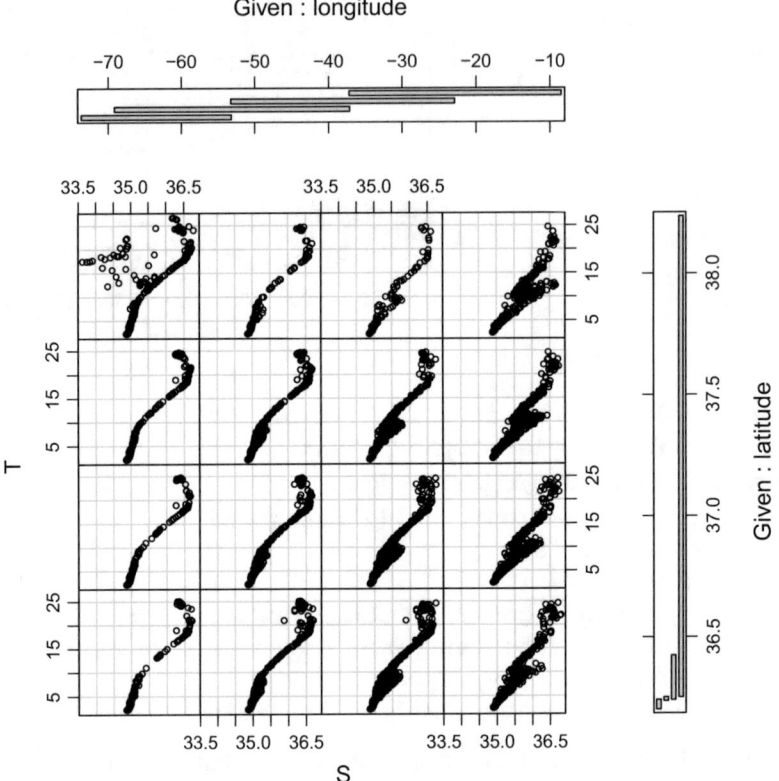

Fig. 6.21 Coplot of spatial $T-S$ relationship in the `section` dataset

```
T <- sec[["temperature"]]
latitude <- sec[["latitude"]]
longitude <- sec[["longitude"]]
```
yields data that are plotted in Fig. 6.21 with
```
coplot(T~S|longitude+latitude, number=4, cex=2/3)
```
where the number of panels has been reduced from the default of 6 to illustrate the broad features. The section is nearly zonal, with little variation in latitude. The scattered points at relatively low salinity and middling temperatures are associated with a location near the coast of North America (see Fig. 5.1, which focusses in that region), and Mediterranean Water is visible as a relatively warm and salty watermass, most prominent towards the east.

Exercise 5.9 on page 135. Suggest how to produce $T-S$ diagrams categorized by longitude, using `cut`, `factor` and `split`.
Solution. Longitude ranges between 68 W and 9 W, and subdivision of stations into 10-degree longitudes is accomplished with

```
cs <- cut(open[["longitude", "byStation"]],
seq(-70, 0, 10))
S <- split(open[["station"]], factor(cs))
```

After which S is a vector of length 7, with each element in turn containing a list of ctd objects, one per station.

A summary plot (not reproduced here) is obtained with

```
plotTS(open, eos="gsw")
```

and aggregate diagrams for each longitude band with

```
for (Si in S) {
    plotTS(as.section(Si), eos="gsw")
    lonm <- mean(unlist(lapply(Si,
                        function(x) x[["longitude"]])))
    mtext(round(lonm, 1), side=3, line=0, adj=1,
    cex=0.8)
}
```

where lapply() has been used to perform a computation across the stations within each longitude band.

Exercise 5.10 on page 136. Formulate a model in which the misfit in S is minimized, and evaluate the confidence interval for it.

Solution. One may begin as in the text with

```
library(deSolve)
eos <- "unesco" # match Schmitt's variables
data(schmitt, package="ocedata")
```

and formulate the model similarly

```
dS.dtheta <- function(theta, S, parms) {
    list(swAlpha(S, theta, 0, eos=eos) /
        parms$Rrho/swBeta(S, theta, 0, eos=eos))
}
SModel <- function(S0, Rrho)
  lsoda(S0,schmitt$theta,dS.dtheta,
  parms=list(Rrho=Rrho))[,2]
fit <- nls(S~SModel(S0, Rrho), data=schmitt,
            start=list(S0=20,Rrho=1))
confint(fit)
            2.5%       97.5%
S0    35.109322 35.150613
Rrho  1.911916  1.969793
```

Exercise 5.11 on page 141. Compare the spring-neap variation in Halifax sea level with the phase of the moon.

Solution. First, load the data and extract relevant variables with

```
data(sealevel, package="oce")
start <- ISOdatetime(2003,3,1,0,0,0,tz="UTC")
end <- ISOdatetime(2003,3,28,0,0,0,tz="UTC")
```

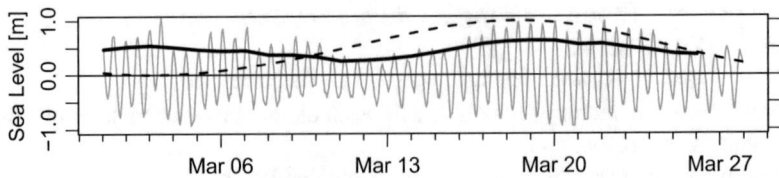

Fig. 6.22 Spring-neap variation of Halifax Harbour sea level, showing the demeaned observed record (gray), an envelop using a running boxcar filter of the squared square (solid black), and the square of fractional area of the lit moon (dashed black)

```
sl <- subset(sealevel, start < time & time < end)
t <- sl[["time"]]
eta <- sl[["elevation"]] - mean(sl[["elevation"]],
na.rm=TRUE)
```
after which it is simple to plot the timeseries $\eta = \eta(t)$ as Fig. 6.22 with
```
oce.plot.ts(t, eta, drawTimeRange=FALSE, col="gray",
            ylim=c(-1,1),
            xlab="", ylab="Sea Level [m]")
abline(h=0)
```
The envelop may be calculated by smoothing the squared signal
```
tt <- as.numeric(t)
env <- binMean1D(tt, eta^2, seq(tt[1], tail(tt, 1),
24*3600))
lines(env$xmids+(t[1]-env$xmids[1]),
sqrt(env$result), lwd=3)
```
Moon phase is given by moonAngle(), but it is perhaps simpler to display the fractional area of the illuminated portion of the moon on the graph.
```
m <- moonAngle(t, 44.65, -63.6)
lines(t, illum <- m$illuminatedFraction, lwd=2, lty=2)
```
Assuming negligible "age of the tide", elevation amplitudes should be highest with full moon and new moons, i.e. illumination fractions near 0 and 1.

Exercise 5.12 on page 142. Use the oce function plotTaylor() to contrast three tidal models of sea level in Halifax Harbour: one with default constituents, one with just M2, and one with just S2.
Solution. After loading the data
```
data(sealevel, package="oce")
eta <- sealevel[["elevation"]]
```
the three tidal fits are constructed and illustrated in Fig. 6.23 with
```
all <- predict(tidem(sealevel))
M2 <- predict(tidem(sealevel, constituents="M2"))
S2 <- predict(tidem(sealevel, constituents="S2"))
plotTaylor(eta, cbind(all, M2, S2))
```

Fig. 6.23 Diagnostic plots of
three tidal models
(Exercise 5.12)

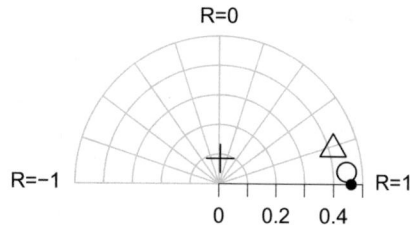

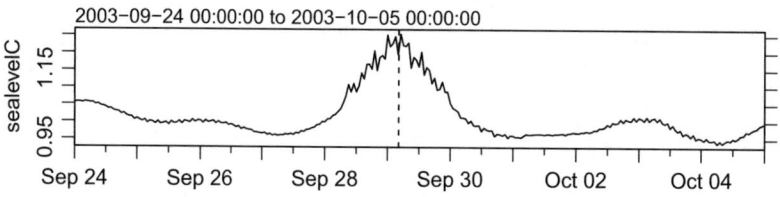

Fig. 6.24 Halifax sea level, after application of a Doodson X0 filter; compare Fig. 5.12

Exercise 5.13 on page 142. Determine whether the storm surge from Hurricane
Juan in Halifax Harbour can be detected after removing tidal energy with a Doodson
tidal filter.

Solution. The Doodson tidal filter is a convolution filter with zero multiplier at zero
lag, and with single-sided multipliers

```
a <- c(2,1,1,2,0,1,1,0,2,0,1,1,0,1,0,0,1,0,1)/30
```

at lags ± 1 h, ± 2 h, etc. (Intergovermental Oceanographic Commission 1985,
Appendix 3), so a full filter is

```
f <- c(rev(a), 0, a)
```

The de-tided signal can be calculated and plotted as in Fig. 6.24 with

```
data(sealevel, package="oce")
eta <- sealevel[["elevation"]]
t <- sealevel[["time"]]
tlim <- as.POSIXct(c("2003-09-24", "2003-10-05"),
tz="UTC")
sealevelC <- filter(eta, f)
oce.plot.ts(t, sealevelC, xlim=tlim, xaxs="i")
abline(v=as.POSIXct("2003-09-29 04:15:00",tz="UTC"),
Slty=2)
```

The results are not encouraging, a comparison with Fig. 5.12 revealing that the
hours-long storm surge event is smeared out over days by the Doodson filter,
decreasing in magnitude by a factor of 5.

Exercise 5.14 on page 142. Construct a hypsometric curve using outer() to
create an area matrix that pairs with the depth matrix topoWorld[["z"]].

Solution. First, construct a percent-area matrix

```
data(topoWorld, package="oce")
z <- topoWorld[["z"]]
```

Fig. 6.25 Hypsometric curve (Exercise 5.14)

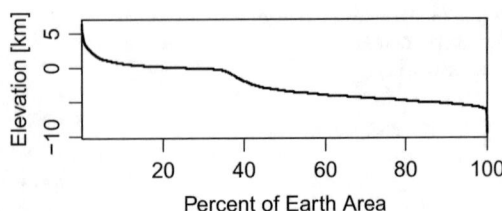

```
lat <- topoWorld[["latitude"]]
area <- outer(rep(1, dim(z)[1]), cos(lat * pi / 180))
area <- 100 * area / sum(area)
```
and then display elevation and fractional area in Fig. 6.25.

```
n <- 100
Z <- seq(min(z), max(z), length.out=n)
a <- unlist(lapply(Z, function(Z) sum(area[z>=Z])))
plot(a, Z/1000, type="s", xaxs="i", lwd=3,
     ylab="Elevation [km]", xlab="Percent of
     Earth Area")
```

Exercise 5.15 on page 143. Graph surface water properties for the delayed-mode portion of the `argo` dataset, as a T–S diagram and a trajectory with symbol size proportional to spiciness.

Solution. The data may be loaded and subsetted with

```
data(argo, package="oce")
argoD <- subset(argo, dataMode=="D")
```
and then near-surface temperature and salinity can be selected with

```
SSS <- apply(argoD[["salinity"]], 2, head, 1)
SST <- apply(argoD[["temperature"]], 2, head, 1)
lon <- argoD[["longitude"]]
lat <- argoD[["latitude"]]
```
after which a CTD object created with

```
SSctd <- as.ctd(SSS, SST, 0, longitude=lon,
latitude=lat)
```
yields the desired spiciness with

```
SSspice <- SSctd[["spice"]]
```
so that Fig. 6.26 can be constructed with

```
par(mfrow=c(1,2))
cexLon <- rescale(lon, rlow=1/2, rhigh=2)
plotTS(SSctd, cex=cexLon)
cexSpice <- rescale(SSspice, rlow=1/2, rhigh=2)
plot(argo, which=1, coastline="coastlineWorld",
cex=cexSpice)
```

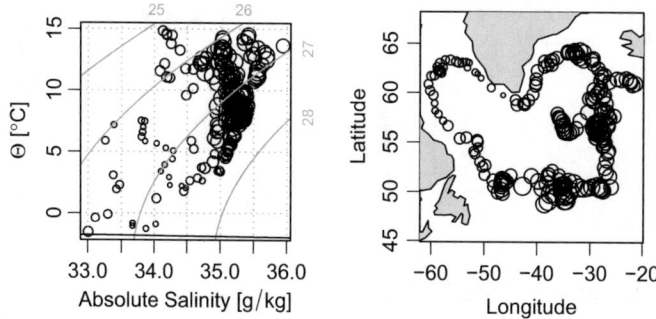

Fig. 6.26 Watermass analysis for the `argo` data set (Exercise 5.15). Left: temperature-salinity diagram with larger symbols for more eastern locations. Right: trajectory of argo surface position, with larger symbols for waters of higher spiciness

Fig. 6.27 Separation of the at-satellite brightness temperature histogram into Gaussian groups. The vertical line is T_B inferred from the warmer group

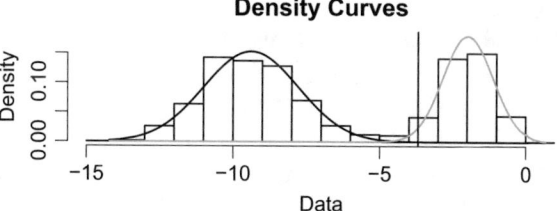

Exercise 5.16 on page 145. Use functions in the `mixtools` package to separate the histogram of Fig. 5.15 into two parts.
Solution. With TB from Sect. 5.8, using

```
library(mixtools)
m <- normalmixEM(TB[!is.na(TB)])
plot(m, which=2, col2=c("black", "darkgray"))
TBice <- m$mu[2] - 2 * m$sigma[2]
abline(v=TBice)
```

yields Fig. 6.27. Note that the inferred brightness temperature of ice, $T_{Bice} = -3.7°C$, is similar to that inferred visually and drawn in Fig. 5.15.

Exercise 5.17 on page 149. Suggest a procedure for performing CTD outlier rejection using data flags, instead of modifying the data.
Solution. With `ctdRaw` loaded and `bad()` defined as in the text, a solution using the World Hydrographic Program flag notation is

```
data(ctdRaw, package="oce")
for (i in 1:2) {
    b <- bad(ctdRaw[["salinity"]])|bad(ctdRaw
    [["temperature"]])
    ctdRaw[["flags"]]$salinity <- ifelse(b, 3, 2)
    ctdRaw[["flags"]]$temperature <- ifelse(b, 3, 2)
```

Fig. 6.28 Comparison of two summaries of log-normal distributions (Exercise 5.18)

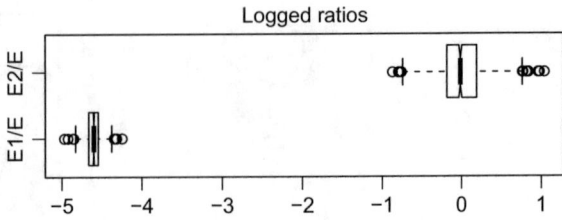

```
        ctdRaw <- handleFlags(ctdRaw, 3)
    }
```
where `handleFlags()` sets the questionable values to `NA` at each of two iterations.

Exercise 5.18 on page 151. Compare E_1 and E_2 in a series of random trials.
Solution.
A function returning E_1/E and E_2/E is
```
    test <- function(n=500,lmuHat=-5*log(10),
    lsigmaHat=2*log(10))
    {
        e <- rlnorm(n, lmuHat, lsigmaHat)
        E <- exp(lmuHat + lsigmaHat^2/2)
        E1 <- exp(mean(log(e)))
        E2 <- exp(mean(log(e)) + 0.5 * sd(log(e))^2)
        c(E1 / E, E2 / E)
    }
```
after which trials may be done with `laply()` in the `plyr` package
```
    library(plyr)
    set.seed(518)                        # for reproducibility
    x <- laply(1:1000, function(trial) test())
```
and the results plotted as a log-space box plot in Fig. 6.28 with
```
    boxplot(log10(x), horizontal=TRUE, notch=TRUE,
            names=c("E1/E", "E2/E"))
    mtext("Logged ratios", side=3, line=0)
```
which shows close agreement between E_2 and E, also evidenced by the 95% confidence interval
```
    10^as.numeric(t.test(log10(x[,2]))$conf.int)
    [1] 0.9632924 1.0462378
```

Exercise 5.19 on page 153. Use `approx` to write a function that interpolates from one set of times to another.
Solution. This solution can be handy for putting a suite of instruments on a common timebase.
```
    adjustTime <- function(t, x, tout, ...)
    {
```

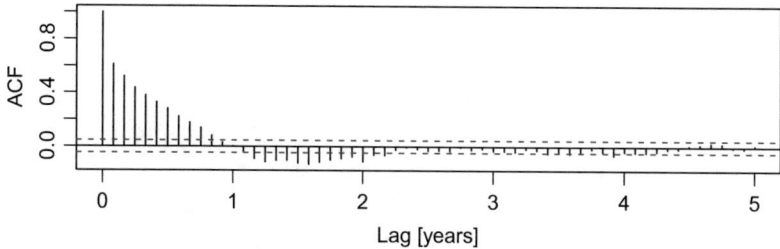

Fig. 6.29 Autocorrelation of `soi` dataset, with confidence intervals

```
    if (missing(tout))
        tout <- seq(min(t), max(t),
        length.out=length(t))
    approx(t, x, tout, ...)$y
}
```

Exercise 5.20 on page 155. Add confidence limits to Fig. 5.19.
Solution. A 95% confidence band may be constructed from $z^*\sigma/\sqrt{n}$, with z^* computed for a t distribution with the number of points in the segment.

```
library(plyr)                              # for laply
n <- laply(is, length)                     # segment lengths
zstar <- qt(1-0.05/2, n)                   # " z values
pm <- zstar * laply(is, sd) / sqrt(n)      # " CI/2
lines(ymean, imean + pm)                   # upper CI range
lines(ymean, imean - pm)                   # lower CI range
```

Exercise 5.21   on page 156. Following Sect. 2.4.8, use `acf()` to look for oscillations in the `soi` dataset.
Solution. Constructing Fig. 6.29 is simple (with the default lag range being extended to show results at up to 5 years):

```
data(soi, package="ocedata")
SOI <- ts(soi$index,
          start=soi$year[1], deltat=diff
          (soi$year[1:2]))
acf(SOI, lag.max=5*12, xlab="Lag [years]", main="")
```

The correlation crosses zero at a 1-year lag, reaches a minimum at approximately 1.5 years, and fails the indicated significance criterion a year thereafter. Doubling the lag at the negative lobe might suggest an oscillation of 3 years, but the low significance in that lag range suggests that any such variations are weak or sporadic.

Exercise 5.22 on page 156. With the `giss` dataset, show how detrending can reduce spurious endpoint effects of Butterworth filters.
Solution. First, plot the data as the gray line in Fig. 6.30

Fig. 6.30 Demonstration of the benefit of detrending before Butterworth lowpass filtering, with gray showing the giss dataset, dashed black showing the result of simple filtering, and solid black showing the result of filtering after detrending (Exercise 5.22)

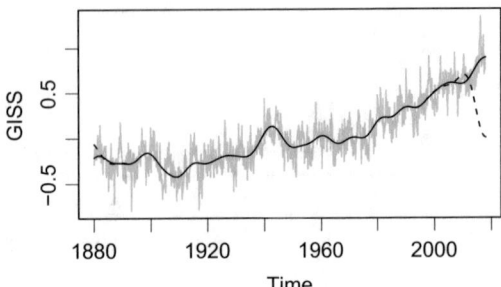

```
data(giss, package="ocedata")
GISS <- ts(giss$index, start=giss$year[1],
frequency=12)
plot(GISS, col="gray")
```
Next, construct a Butterworth filter with half-power at 120 months (quarter-power, with filtfilt() doing the filtering)
```
library(signal)
f <- butter(n=4, W=2*1/120)      # W=2*tauS/tauC in
                                   months
```
Applying this filter and drawing the results on Fig. 6.30
```
A <- filtfilt(f, GISS)
lines(giss$year, A, lty=2, lwd=2)
```
illustrates how Butterworth filters can produce spurious effects near the endpoints. Better results may be achieved by first detrending the time series, pulling the endpoints towards zero.

A simple way to remove a trend is with detrend() in the oce package, e.g.
```
dt <- detrend(giss$index)
```
creates a list containing Y, the detrended values, along with a and b, the slope and intercept with respect to seq_along(giss$index).

Filtering the detrended time series, adding back the removed trend, and plotting as a solid curve in Fig. 6.30
```
B0 <- filtfilt(f, dt$Y)
B <- B0 + dt$a + dt$b * seq_along(giss$year)
lines(giss$year, B, lwd=2)
```
reveals that the two filtering schemes yield similar results in the middle years of observation, but that detrending produces results that are more faithful to the data at the start and (particularly) the end.

Exercise 5.23 on page 160. Use spectrum() to demonstrate the demonstrate Butterworth filter response.
Solution. It is convenient to use rnorm() to construct a time series
```
x <- rnorm(1e5)
```

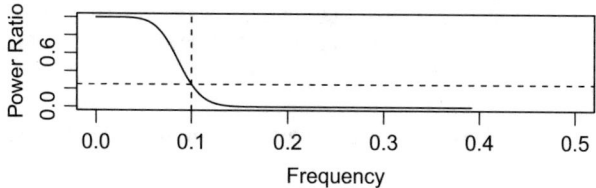

Fig. 6.31 Frequency response of third-order Butterworth filter with cutoff 0.5 times the Nyquist frequency, applied with `filtfilt()` (Exercise 5.23)

after which a third-order Butterworth filter with 3-dB cutoff at, say, one-fifth the Nyquist frequency may be constructed with

```
library(signal)
W <- 1 / 5
filter <- butter(3, W)
```

It is recommended to use `filtfilt()` to avoid altering the phase

```
y <- filtfilt(filter, x)
```

but that will change the half-power point to a quarter-power point, as illustrated in Fig. 6.31 with the following.

```
X <- spectrum(x, plot=FALSE)
Y <- spectrum(y, plot=FALSE)
plot(X$freq, Y$spec / X$spec, type="l",
     xlab="Frequency", ylab="Power Ratio")
abline(v=W*0.5, lty="dashed") # Nyquist=0.5
abline(h=1/4, lty="dashed")
```

Exercise 5.24 on page 161. Write a function to compute and display a frequency-time plot.

Solution. A simple solution is as follows.[7]

```
sonogram <- function(t, x, seglen=64, log=FALSE, ...)
{
    n <- length(t)
    if (length(x) != n) stop("t and x must be of equal length")
    n <- n - n %% seglen # for trimming to multiple of seglen
    t <- t[1:n]
    x <- x[1:n]
    xs <- split(x, cut(t, breaks=floor(n/seglen)))
    m <- length(xs[[1]])
    freq <- spectrum(xs[[1]], plot=FALSE)$freq
    nfreq <- length(freq)
    spec <- matrix(NA, nrow=length(xs), ncol=nfreq)
    for (i in seq_along(xs))
        spec[i,] <- spectrum(xs[[i]], plot=FALSE)$spec
    spec <- spec / sqrt(nfreq) # normalize as for spectrum()
```

---

[7]Note the normalization of the `fft()` output, to match the convention of `spectrum()`.

```
        time <- t[seq(1, by=m, length.out=length(xs))]
        imagep(time, freq, if (log) log10(spec) else spec,
            xlab="Time", ylab="Frequency", ...)
}
```

where the %% operator is one of several ways of trimming one interval to a multiple
of another integer.

This function may be tested with a chirp signal

```
n <- 3600
t <- 1:n
freq <- 2 * pi * (0.1 + 0.2 * t/n)
x <- rnorm(n) + sin(freq*t)
sonogram(t, x, col=function(n) gray.colors(n, 1, 0))
```

with results as shown in Fig. 5.23 on page 159.

Exercise 5.25 on page 161. Use fft() to compute rotary spectra as defined by
Gonella (1972).

Solution. With time-series u and v, Gonella (1972) writes the clockwise and
anticlockwise rotary spectra as

$$S_- = \frac{1}{8} (P_{uu} + P_{vv} - 2Q_{uv})$$

$$(6.7)$$

$$S_+ = \frac{1}{8} (P_{uu} + P_{vv} + 2Q_{uv})$$

where P_{uu} and P_{vv} are the auto-spectra of u and v, and Q_{uv} is the quadrature
spectrum between u and v. This may be expressed as[8]

```
spec.rotary <- function(u, v, deltat=1)
{
    n <- length(u)
    U <- fft(u) / sqrt(n)
    V <- fft(v) / sqrt(n)
    Puu <- U * Conj(U)
    Pvv <- V * Conj(V)
    Quv <- -Re(U)*Im(V) + Re(V)*Im(U)
    cw  <- (Puu + Pvv - 2*Quv) / 8
    acw <- (Puu + Pvv + 2*Quv) / 8
    keep <- seq(2, floor(n/2) + 1)
    f0 <- 1 / deltat
    f <- seq(f0/n, by=f0/n, length.out=floor(n/2))
    list(freq=f, specCW=Re(cw[keep]),
            specCCW=Re(acw[keep]))
}
```

---

[8]As in the previous Exercise, note the scaling of fft() output.

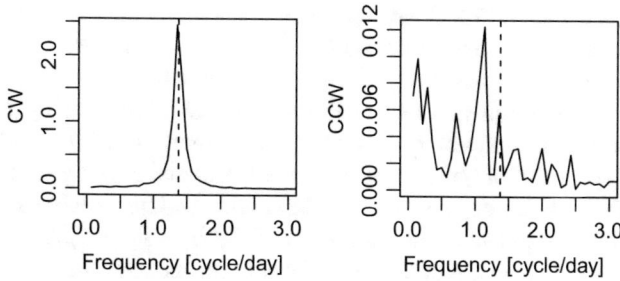

Fig. 6.32 Rotary spectra of velocities predicted with an inertial-oscillation model, showing (left) clockwise and (right) anticlockwise power, focussed on low frequencies including (dashed) the inertial. Note the 200-fold scale difference (Exercise 5.25)

As a test, u and v time series were constructed with an inertial model similar to that in Sect. 5.9.6, with altered forcing and added noise. As Fig. 6.32 shows, the inertial-band energy is mostly confined to the clockwise spectrum, as expected for this northern-hemisphere simulation.

Exercise 5.26 on page 163. Use na.kalman from the imputeTS package to replace the spikes in x with the predictions of a Kalman filter.
Solution. With x and bad() defined as in the text,

```
y <- x
y[bad] <- NA
library(imputeTS)
yy <- na.kalman(y)
plot(t, x, type="l")
lines(t, yy, col="red")
```

will define y, with NAs in the spikes, and yy, with the NAs replaced with the predictions of a Kalman filter. This call to na.kalman() uses a maximum-likelihood model, but other models are also possible (Moritz and Bartz-Beielstein 2017).

Exercise 5.27 on page 168. Use the zd element of the return value from interpBarnes() to add the percentage errors to Fig. 5.28.
Solution. This misfit may be added with

```
library(plyr)
e <- laply(seq_along(topo$x),
          function(i) (topo$z[i]-b$zd[i])/topo$z[i])
text(topo$x, topo$y, round(100 * e), cex=0.9)
```

Exercise 5.28 on page 172. Construct a two-layer box model for temperature, with the top layer subjected to a sinusoidal heat flux. Assume the boxes to be of equal thickness, and devise a convection scheme to prevent temperature inversion.

Solution. The method involves a function defining dT_1/dt and dT_2/dt, where T_1 and T_2 are the temperatures in the upper and lower boxes, and t is time. Take the heat flux to be $F = A \sin(2\pi t/\tau)$, where A is a constant and $\tau = 1$ year. Let the layer depths be H, the water density be $\rho = 1027 \text{ kg/m}^3$ and the specific heat be $C_P = 4300 \text{ J/(kg °C)}$. The model equations are then $dT_1/dt = F/(\rho C_P H) - k(T_1 - T_2)$ and $dT_2/dt = k(T_1 - T_2)$. A crude way to handle convection is to increase k greatly when $T_1 < T_2$.

```
model <- function(t, y, parms, ...)
{
    T1 <- y[1] # extract elements for clarity of notation
    T2 <- y[2]
    F <- parms$A * sin(2 * pi * t / parms$tau)
    K <- if (T1 < T2) 1e3 * parms$k else parms$k
    list(c(dT1dt = F / (1027*4300*parms$H) - K * (T1 - T2),
          dT2dt = K * (T1 - T2)))
}
```

The solution for initial condition $T_1 = T_2 = 10°C$, given parameter values $A = 200 \text{ W/m}^2$, $\tau = 1\text{year}$, $k = 1 \times 10^{-7} \text{ s}^{-1}$ and $H = 50 \text{ m}$, is found with

```
library(deSolve)
IC <- c(10, 10)
spy <- 365 * 86400                        # seconds per year
parms <- list(A=200, k=1e-7, H=50, tau=1 * spy)
times <- seq(0, 2 * spy, length.out=500)
T12 <- lsoda(IC, times, model, parms)
```

and the resultant Fig. 6.33 is created with

```
plot(T12[,1]/spy, T12[,2], type="l",
     xlab="Year", ylab="Temperature")
lines(T12[,1]/spy, T12[,3], lty=2)
```

Exercise 5.29 on page 172. Develop a numerical solution to the convection problem formulated by Stommel (1961), $dy/dt = 1 - y - (y/\lambda)|Rx - y|$ with $dx/dt =$

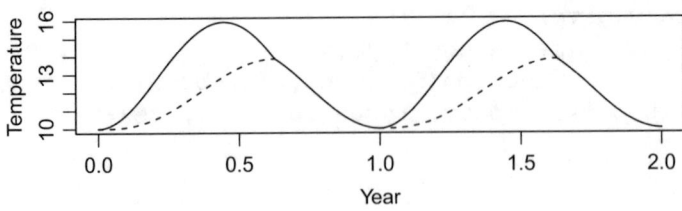

Fig. 6.33 Evolution of temperatures in a two-box model of convection, with solid and dashed lines for upper and lower layers (Exercise 5.28)

$\delta(1 - x) - (x/\lambda)|Rx - y|$, where x is dimensionless salinity, y is dimensionless temperature, t is dimensionless time, $\delta = 1/6$, $\lambda = 1/5$ and $R = 2$. Draw some traces to mimic Stommel's Figure 7, including one starting at $x = 0.55$, $y = 1$, which approaches a stable-spiral attractor.

Solution. The differential equations are returned by the function

```
DE <- function(t, xy, param, ...) {
    x <- xy[1]                         # shorthand for clarity
    y <- xy[2]
    list(c(param$delta*(1-x)-(x/param$lambda)*abs(param$R*x-y),
        1-y-y/param$lambda*(abs(param$R*x-y)))))
}
```

With parameters as used by Stommel (1961)

```
param <- list(R=2, delta=1/6, lambda=1/5)
```

and report times

```
times <- seq(0, 20, length.out=200)
```

the calculation and plotting are handled with

```
library(deSolve)
soln <- lsoda(y=c(0.55, 1), times=times, func=DE,
param=param)
plot(soln[,2], soln[,3], type="l",
    xlab="x", ylab="y", xaxs="i", yaxs="i",
    xlim=c(0, 1), ylim=c(0, 1))
```

which starts Fig. 6.34, the other trajectories added to which use starting points chosen to mimic Stommel's values. To guide the eye to the spiral attractor (and another attractor), a filled symbol indicates the start of each trajectory and an open symbol indicates the end.

Exercise 5.30 on page 177. In her 1972 song "You're so vain," Carly Simon mentions flying to Nova Scotia to view a solar eclipse. Determine the time of that eclipse, assuming that it occurred on March 7, 1970. Use `optimize()` with the oce functions `moonAngle()` and `sunAngle()`.

Fig. 6.34 Evolution of nondimensional temperature (x) and salinity (y) in the Stommel (1961) convection model (Exercise 5.29)

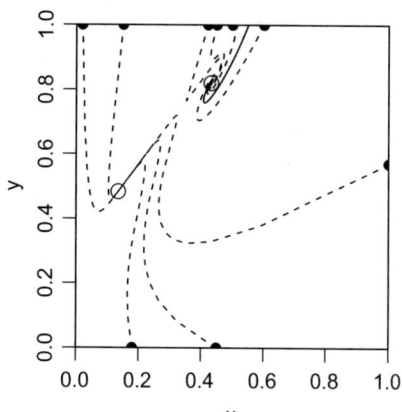

Solution. The first step is to create a function returning a measure of the sun-moon separation at (for example) the location of Halifax. One method is to sum squared deviations in altitude and azimuth.

```
lat <- 44.6500
lon <- -63.6000
distance <- function(t)
{
    t <- numberAsPOSIXct(t)
    m <- moonAngle(t, lat, lon)
    s <- sunAngle(t, lat, lon)
    sqrt((m$azimuth - s$azimuth)^2 +
         (m$altitude - s$altitude)^2)
}
```

Next, define limits for the day in question

```
l <- as.POSIXct("1970-03-07 00:00:00", tz="UTC")
u <- l + 86400
```

Now, use `optimize()` to perform the 1-D optimization.

```
eclipse <- optimize(distance, lower=l, upper=u)
```

The function returns its smallest value at time

```
numberAsPOSIXct(eclipse$minimum)
```
```
[1] "1970-03-07 17:39:55.204 UTC"
```

and that value is

```
eclipse$objective
```
```
[1] 0.4540549
```

The moon and sun each cover about 0.5 degrees of the sky, so this is indeed an eclipse. Readers might enjoy determining whether this was a *total* eclipse somewhere in Nova Scotia, as Simon indicates. As for that famous question relating to this song, R has little to say.

Exercise 5.31 on page 181. Use `svd()` to apply the singular value decomposition method to the `adp` dataset.

Solution. Reconstruct the data as in the text, combining a few of the steps.

```
data(adp, package="oce")
adp2 <- subset(adp, distance < 38)
u <- adp2[["v"]][,,1]
distance <- adp2[["distance"]]
time <- adp2[["time"]]
```

Missing values cannot be handled by `svd()`, but replacing them with the median may be reasonable

```
u[is.na(u)] <- median(u, na.rm=TRUE)
```

after which the actual decomposition is a trivial function call

```
U <- svd(u)
```

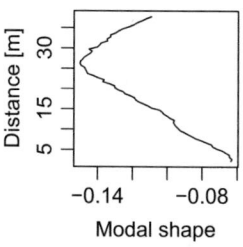
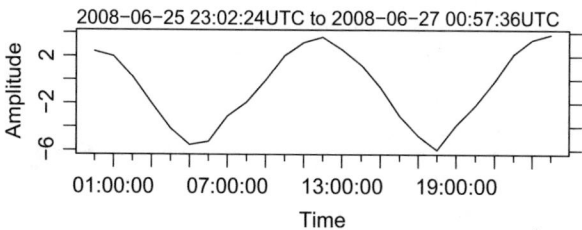

Fig. 6.35 Empirical orthogonal function analysis of ADCP data (Exercise 5.31)

The modal shapes are in the columns of U$u, while the time-varying amplitudes are in the columns of U$v. The results for the first mode may be plotted as Fig. 6.35 with the following, where layout () is used to set up panels of unequal width.

```
layout(matrix(1:2, nrow=1), widths=c(0.3, 0.7))
plot(-U$v[,1], distance, type="l",
        xlab="Modal shape", ylab="Distance [m]")
oce.plot.ts(time, -U$d[1]*U$u[,1],
                xlab="Time", ylab="Amplitude")
```

Exercise 5.32 on page 183. Explore the accuracy of this method by adjusting Δz with the case of constant $N = 0.01 \, \mathrm{s}^{-1}$.
Solution. The theoretical wave speed $\sqrt{N H}/\pi = 0.3183$ m/s for $N = 0.01 \, \mathrm{s}^{-1}$ and $H = 100$ m. Tests with code as in the text recover this value with absolute errors of 0.5, 0.98 and 1.92% for $\Delta z = 1, 2$ and 4 m; all of these values are less than the 5% error expected for a typical 10% uncertainty in N.

Exercise 5.33 on page 186. Create a neural network to describe the variation of sound speed with pressure, for station 103 of the section dataset, exploring the effect of variable transformation on reproducibility.
Solution. First, load and plot the data for this station, which is in 5000 m of water, south of Cape Cod at 36 °N.

```
library(nnet)
data(section, package="oce")
stn <- handleFlags(section[["station", "104"]])
par(mfrow=c(1, 2))
speed <- swSoundSpeed(stn)                    # in m/s
p <- stn[["pressure"]]                         # in dbar
plotProfile(stn, xtype=speed, xlab="Speed [m/s]",
type="p")
```

The resultant circles in Fig. 6.36 reveal a SOFAR channel centred at about 1500 m. The scales of sound speed, pressure, and the derivative of the former with respect to the latter are very dissimilar, which raises concerns about the ability of the optimizer used by nnet () to find a solution. Since that optimizer starts

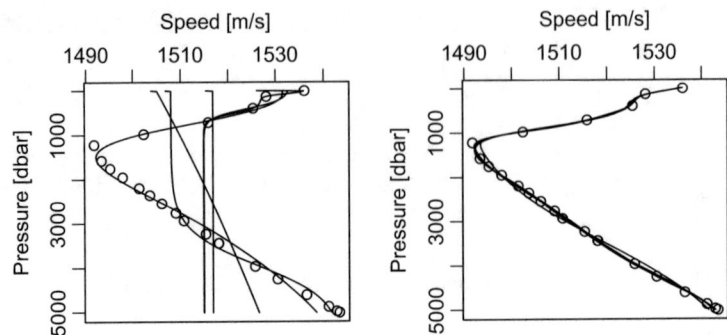

Fig. 6.36 Neural network model of sound speed as a function of pressure at station 103 of the section dataset. Left: data and the predictions of 10 trials with unscaled units. Right: similar, but with transformed variables (Exercise 5.33)

with a random number, a crude way to assess the level of difficulty is to test for reproducibility across repeated trials.

For illustration, the solutions will be shown on a uniform pressure grid

```
set.seed(533)                              # for reproducibility
pp <- seq(max(p), 0, -10)
trials <- 10
e1 <- NULL # will hold misfits
for (i in 1:trials) {
    n1 <- nnet(speed ~ p, linout=TRUE,
                size=5, decay=1e-2, maxit=1000)
    speedPred1 <- predict(n1, newdata=data.frame(p=pp))
    lines(speedPred1, pp)
    e1[i] <- mean(abs(speed - predict(n1,p)))
}
```

The results as drawn in Fig. 6.36 are certainly too variable to be of practical use. One of the trials has predictions that closely follow the data throughout the water column, but the others don't even display a SOFAR channel; indeed, the mean absolute misfit, e1=10.7 m/s, is comparable to the standard deviation of sound speed over the depth (16 m/s).

A reasonable variable transformation might be to express speed as a deviation from a canonical value of 1500 m/s and pressure in 1000 dbar units, since this will pull the variables and their derivatives closer to $\mathcal{O}(1)$. In these new terms, the solution becomes

```
A <- 1500
B <- 1000
Speed <- speed - A
P <- p / B
plotProfile(stn, xtype=speed,xlab="Speed [m/s]", type="p")
e2 <- NULL
for (i in 1:trials) {
    n2 <- nnet(Speed ~ P, linout=TRUE,
                size=5, decay=1e-2, maxit=1000)
```

```
    speedPred2 <- A + predict(n2, newdata=data.frame(P=pp/B))
    lines(speedPred2, pp)
    e2[i] <- mean(abs(speed - (A + predict(n2,p))))
}
```

which results in the right panel of Fig. 6.36. The improvement is obvious: the transformed variables yield results that describe the data very closely. Indeed, the misfit e2=0.4 m/s for the transformed variables is low enough that an analyst might wonder whether overfitting will be a problem. A natural step in exploring that idea might involve exploring the results of varying the decay argument of nnet(). Another approach would be to apply cross validation methods to the problem, perhaps along lines discussed by Venables and Ripley (1999).

Since neural networks can display surprisingly odd results at times (see, e.g., Hutson 2018), practical analysts tend to spend a great deal of time testing the robustness and extendibility of proposed solutions. They also sense become familiar with the relative merits of the different approaches to neural networks that are provided in R, since nnet() is certainly not the only choice. As noted in the text, neural networks are a powerful tool, but not one to be used blindly.

Appendix A
Switching from Matlab to R

Most analysts find that it is not difficult to switch from Matlab to R, especially if a few key differences are kept in mind. This appendix provides a list of such differences, gleaned from the experience of the author and his colleagues (especially Clark Richards, who helped to compile the list).

Syntax

1. Assignment to a variable is denoted with "=" in Matlab, but with "<-" in R. (Actually, it *is* possible to use "=" for R assignment, but not recommended.)
2. In R, assignment statements do not print their value, so there is no need for the Matlab convention of using ";" for silencing an assignment.
3. In R, as in most modern languages *except* Matlab, square brackets are used for indexing; see Sect. 2.3.4.
4. R matrices are not constructed with a square-bracket syntax, but rather with `matrix()`, `as.matrix()`, `cbind()` or `rbind()`; see Sect. 2.3.5.
5. In Matlab, vectors (one-dimensional sequences of values) are often represented as single-column matrices. The same form can be used in R, but most functions work with vectors, instead. The `drop()` function, which drops unused matrix dimensions, helps to convert Matlab data to R format, e.g. the following shows how to create vectors for regression with `lm()`.

```
library(R.matlab)
m <- readMat("filename.mat")
x <- drop(m$x)
y <- drop(m$y)
lm(y ~ x)
```

6. R coerces arrays to a lower dimension whenever it can (Sect. 2.3.5.2), but the `drop=FALSE` argument can override this, e.g.

© Springer Science+Business Media, LLC, part of Springer Nature 2018
D. E. Kelley, *Oceanographic Analysis with R*,
https://doi.org/10.1007/978-1-4939-8844-0

```
a <- matrix(1:10, nrow=2)
a[1,]                                              # a vector
[1] 1 3 5 7 9
```

yields a vector, whereas

```
a[1,,drop=FALSE]                                   # a 1-row matrix
      [,1] [,2] [,3] [,4] [,5]
[1,]    1    3    5    7    9
```

yields a one-row matrix. (Note the second comma in the drop=FALSE case.)

7. Matrix multiplication in R uses the %*% operator, while the * operator does item-by-item multiplication. See Sect. 2.3.5.1 for this and other matrix operations.

8. In Matlab, a period in a variable name indicates the selection of a subcomponent of a structure. In R, a period in a variable name is generally taken to have no particular meaning *except* for generalized functions. See Sect. 2.3.2.

9. The q() function is called to exit R. Since this is a function call, the parentheses are required. Dropping the parentheses yields a cryptic message that does little to suggest that R is a friendly language!

Graphics

1. In Matlab, hold on is used to indicate a desire to embellish an existing plot. Instead, R provides a suite of functions whose whole purpose is to add to existing plots, such as points() for adding points, lines() for adding lines, title() for adding titles, legend() for adding legends, mtext() for writing in plot margins, etc., plus an add argument to contour() and a few other functions to make them add to an existing plot; see Sect. 2.4.

2. Matlab offers better interactive control of plots than is available in the basic R graphics system, although the shiny package makes up for this, at some coding cost. See Sects. 2.4.15 and 2.8, plus Appendix B.

3. Matlab produces "flashier" default graphics, e.g. automatically using colours to distinguish between lines on a plot. The R strategy is to produce more utilitarian black/white default plots, in accordance with the tenets outlined by Cleveland and McGill (1984) and others who have studied the interpretation of graphical material. R also offers colour schemes that are suitable for viewers with vision limitations (Ihaka 2003; Light and Bartlein 2004; Zeileis et al. 2009).

Freedom

1. Matlab is a commercial product, sold at a price that is significant to many research groups and is likely to be prohibitive to those "citizen scientists" who might wish to use code provided in the supplemental materials of research papers. R costs nothing.

2. Portions of Matlab are in closed-source form, making it difficult for users to check the methods for veracity or appropriateness. The entire R source is open to inspection.
3. Matlab is covered by commercial licences that may be untoward in some circumstances. R is covered by a GNU license.

Appendix B
GUI Systems for R

Graphical user interfaces (GUIs) simplify the interactive use of R, providing benefits to users of various skill levels. Those with limited programming experience may find GUIs less daunting than the R command-line interface. Those will more programming experience may appreciate the ease with which GUIs let them access previous commands, plots, and documentation views. Even experts may use GUIs when they need simple access to debugging tools, without losing the ability to use the editors and revision-control systems of their choice.

The basic R system centres on a command-line mode, but this is supplemented by various GUI systems that can ease interactive analysis (Fig. B.1). Many users switch between GUI and the command-line, for different sorts of tasks or for a given task at different stages of completion.[1] Some R newcomers, especially those with limited programming experience, focus almost entirely on GUI systems, but even experienced programmers can benefit from GUIs for occasional or everyday use.

There are several GUIs to choose from. Most are in continuous development, making it problematic to provide detailed feature comparisons here. Still, a broad and quite personal overview may be of some use to readers.

- Rstudio is a multi-platform system that is popular across a wide range of R users. Beginners appreciate the access it provides to previously viewed documentation entries and plots, along with its provision of a simple data viewer and a code-aware text editor. Those with more computing skills will appreciate that Rstudio works well with system tools, such as external editors and revision control systems. Developers will appreciate the ease of rebuilding packages,

---

[1] For example, the author tends to use Rstudio to build packages, but not for data analysis. For the latter, he uses Mac-GUI (usually coupled with a Vim editor window) for exploratory work, moving to standalone script files as the work progresses. For any task that takes more than a minute, these script files are run with a Makefile so that analysis is repeated only when the R source-code or the data files are changed.

© Springer Science+Business Media, LLC, part of Springer Nature 2018
D. E. Kelley, *Oceanographic Analysis with R*,
https://doi.org/10.1007/978-1-4939-8844-0

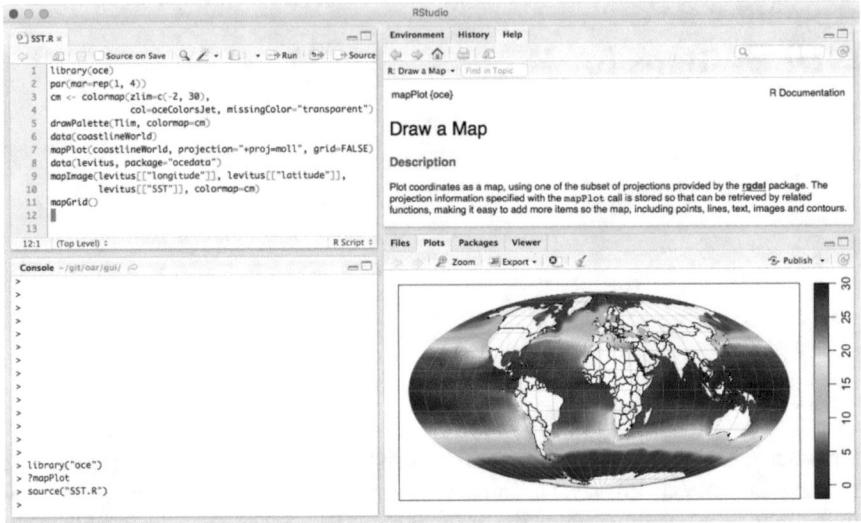

Fig. B.1 An Rstudio window, configured with panels for source code, documentation, console and graphics

interacting with the debugger, running code-quality checkers, etc. Users at all levels can benefit from Rstudio notebooks, which record interactive work in webpages that combine code and graphs.

- Mac-GUI is a Macintosh-only system that provides menu access to documentation, graphs, console, and editor windows. It provides several Rstudio features, although typically in more limited ways. Its restriction to a single platform is a limitation for users who switch between different types of computer. Mac-GUI does not follow the Rstudio tendency of merging windows together into a single frame, so it can be a good choice for those who prefer to size and arrange their own windows.
- JGR supports some of the features of Rstudio, in a more limited way.
- Rcommander provides menu items for some operations such as data loading, regression and plotting, which may be appealing to beginners.

In addition to GUI systems, there are connections between R and several of the general text editors that are popular in technical fields. This is important because users with programming experience tend to have a favourite code-aware text editor with which they are particularly adept. Emacs users can run R with the ESS (Emacs Speaks Statistics) system, while Vim users can use the Vim-R-plugin system. Both of these are set up for R syntax so they can indent and colourize appropriately, and both interact with R to let users execute blocks of code, consult documentation, and perform other tasks involved in using R.

Appendix C
Map Projections in `oce`

The `oce` package provides many of the map projections used in modern cartography. The calculations are done with the `rgdal` package as an interface to the underlying `PROJ.4` system, and this means that the notation will be familiar to readers who have experience with map projections in other computing languages. Only projections that may be inverted are incorporated in `oce`, because inverse calculations are required for labelling axes, etc. A table of the roughly 100 `oce` projections is presented in this appendix, along with general remarks on choosing projections and the details of a few common projections.

Cartesian latitude-longitude plots may be suitable for limited areas, if meridional convergence is handled by setting `asp` to the reciprocal cosine of mean latitude. However, a projection may be preferable for scales exceeding a few hundred kilometres. There are many projections to choose from, and no firm rules to guide the selection. Often, a projection is chosen to match previous work, facilitating visual comparison of the data being displayed. The distortion of shapes and areas are also important considerations (see Airy (1861) for an early treatment and modern discussions by Snyder (1987) and Evenden (2003), the last of which employs the `PROJ.4` notation used in `oce`).

Projections may be classified roughly in three categories: cylindrical, conical, and azimuthal. Conceptually, cylindrical projections have a light source at the earth centre, with rays passing through points of interest on the earth surface and then onto a cylinder enclosing the earth. Such projections are well-suited to equatorial regions. Conical projections have a cone receiving the light, and are often used for mid-latitude regions. Azimuthal projections, in which a tangent plane receives the light, are useful for polar regions. Many variants of these basic projections exist, along with representations that do not fit the light-casting analogy; see Snyder (1993) for an overview and history.

The `oce` package handles map projections with the `rgdal` interface to a system-level `PROJ.4` library.[1] Maps are created with `mapPlot()` and adorned with

---

[1] https://trac.osgeo.org/proj/.

© Springer Science+Business Media, LLC, part of Springer Nature 2018
D. E. Kelley, *Oceanographic Analysis with R*,
https://doi.org/10.1007/978-1-4939-8844-0

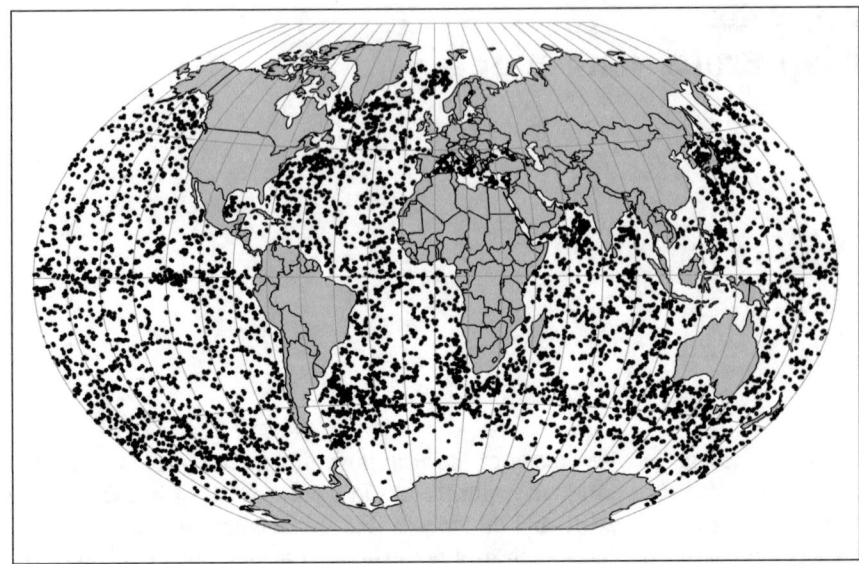

Fig. C.1 World view, using the Winkel Tripel projection, popularized by the National Geographic Society. The dots are the positions of Argo floats in January, 2018. See Page 248 for the code used to create this diagram

mapPoints() for points, etc. A list of oce projections is given in Tables C.1 and C.2, with details stemming from the PROJ.4 documentation. A few of these projections have already been illustrated, including Mercator (page 82), Mollweide (page 99), Lambert Conformal Conic (pages 170 and 203), Robinson (page 188), and Winkel Tripel (page 248). NOAA uses the Robinson projection, as did the National Geographic Society until a recent shift to Winkel Tripel. The latter is illustrated in Fig. C.1, which was created with

```
data(coastlineWorld, package="oce")
mapPlot(coastlineWorld, projection="+proj=wintri",
        col="lightgray")
year <- 2018
month <- 1
url <- "https://data.nodc.noaa.gov/argo/inv/basins"
for (basin in c("atlantic", "pacific", "indian")) {
    f <- sprintf("%s/%s/%s/%s%s%02d_argoinv.txt", url, basin,
                 year, substr(basin, 1, 2), year, month)
    d <- read.csv(f, stringsAsFactors=FALSE)
    mapPoints(d$longitude_min, d$latitude_min,
              pch=20, cex=0.4)
}
```

Fig. C.2 Left: Orthographic projection of the Arctic and North Atlantic. Right: stereographic of the Arctic

In contrast to the Robinson and Mollweide projections, Winkel Tripel curves latitude lines, perhaps clarifying features of the Antarctic coastline.

At the hemispheric scale, it can be useful to use projections that call to mind a view from space, e.g. the left panel of Fig. C.2 is created with an orthographic projection

```
mapPlot(coastlineWorld, projection="+proj=ortho +
        lat_0=60", col="lightgray", drawBox=FALSE)
```

Polar views commonly use a stereographic projection, e.g. the right panel of Fig. C.2 results from

```
mapPlot(coastlineWorld, col="gray", axes=FALSE,
        projection="+proj=stere +lat_0=90 +lon_0=-120",
        longitudelim=c(-180, 180), latitudelim=
        c(70, 110))
```

where lat_0 yields a northern view, lon_0 puts the Beaufort Sea at the bottom of the plot, and the symmetry of latitudelim about 90 centres the pole.

In addition to factors relating to domain size, one must also consider the issue of area and shape distortion. For example, Fig. C.3 illustrates the difference between an Albers equal area projection

```
p <- "+proj=aea +lat_1=15 +lat_2=60 +lon_0=-45"
mapPlot(coastlineWorld, projection=p, col="gray",
        longitudelim=c(-90, 0), latitudelim=c(0, 50))
mapTissot()
```

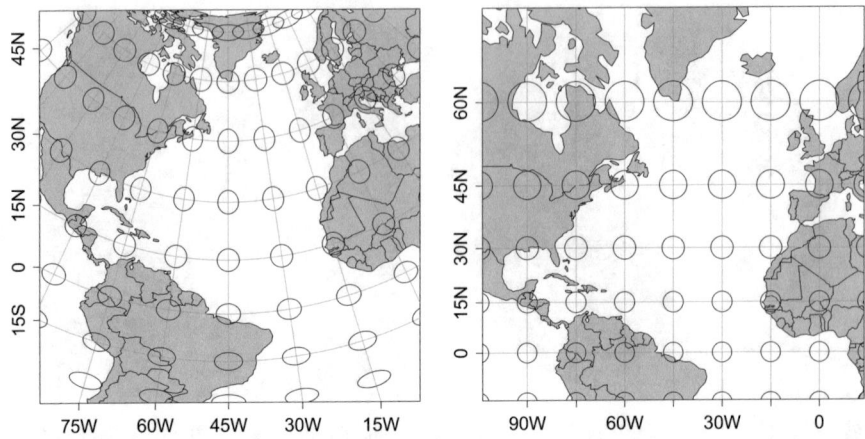

Fig. C.3 Comparison of Albers equal area projection (left) and Mercator projection (right), each with Tissot indicatrices to indicate area and shape distortion

and the Mercator projection

```
p <- "+proj=merc"
mapPlot(coastlineWorld, projection=p, col="gray",
        longitudelim=c(-90, 0), latitudelim=
        c(-10, 70))
mapTissot()
```

revealing that the latter yields noticeable area distortion at high latitudes. Thus, the Albers (or some other area-preserving) projection might be preferred for tasks in which area is of great concern, such as mapping air-sea heat flux. However, the Mercator projection is conformal, so that it preserves shapes, and this can be a prime factor in other applications.

Visual guidance on selecting a projection is provided by mapTissot(), which was used to draw the semi-circular shapes in Fig. C.3. These show how imaginary circles on the earth are transformed by the projection (Tissot 1881; Snyder 1987). In this case, the resultant shapes, a form of Tissot indicatrices, provide a clear display of how the Mercator projection distorts area but not shape, while the Albers projection distorts shape but not area. Note that the illustrated Albers projection has lat_1 and lat_2 set to minimize distortion at 15 and 60°N, yielding relatively low distortion in mid latitudes and higher distortion elsewhere.[2]

Future versions of the oce package will incorporate new versions of PROJ.4 as they become available. This should mean an increase in the number of projections, and a decrease in problems with them. However, the evolution of PROJ.4 may lead

---

[2]This high degree of configuration is a strength of the PROJ.4 projection used by oce, but it also imposes a responsibility on authors to provide more details than is common in the literature, so that readers can reproduce results.

to incompatibilities, if the arguments to projections change (as they did between versions 4.8 and 4.9), so that Tables C.1 and C.2 may go out of date over the years. This should not be of great concern, however, since the documentation for `mapPlot()` will be updated as the capabilities change. Indeed, the assumption is that readers will consult that documentation during the course of their work, e.g. to learn about the available models of the earth ellipse, as well as the various chart-datum schemes that can be used.

Table C.1 Map projections in `oce`

Name	Description	Type	Arguments
aea	Albers Equal Area	Conic Sph&Ell	`lat_1, lat_2`
aeqd	Azimuthal Equidistant	Azi, Sph&Ell	`lat_0, guam`
aitoff	Aitoff	Misc Sph	–
bipc	Bipolar conic, western hemi.	Conic Sph	–
bonne	Bonne (Werner if `lat_1=90`)	Conic Sph&Ell	`lat_1`
cass	Cassini	Cyl, Sph&Ell	–
cc	Central Cylindrical	Cyl, Sph	–
cea	Equal Area Cylindrical	Cyl, Sph&Ell	`lat_ts`
collg	Collignon	PCyl, Sph	–
crast	Craster Parabolic (Putniņš P4)	PCyl, Sph	–
eckN	N=1 to 6 for Eckert I to VI	PCyl, Sph	–
eqc	Equidistant Cylindrical	Cyl, Sph	`lat_ts, lat_0`
eqdc	Equidistant Conic	Conic, Sph&Ell	`lat_1, lat_2`
euler	Euler	Conic, Sph	`lat_1, lat_2`
etmerc	Extended Transv. Mercator	Cyl, Sph	`lat_ts, lat_0`
fahey	Fahey	Pcyl, Sph	–
fouc	Foucault	Pcyl, Sph	–
fouc_s	Foucault Sinusoidal	Pcyl, Sph	–
gall	Gall Stereographic	Cyl, Sph	–
geos	Geostationary Satellite View	Azi, Sph&Ell	h
gn_sinu	General sinusoidal series	Pcyl, Sph	m, n
gnom	Gnomic	Azi, Sph	–
goode	Goode homolosine	Pcyl, Sph	–
hatano	Hatano Asym. Eq. Area	Pcyl Sph	–
healpix	HEALPix	Sph Ellips	–
rhealpix	rHEALPix	Sph Ellips	`north_square, south_square`
igh	Interrupted Goode Homolosine	Pcyl Sph	–
imw_p	Inter. map of world polyconic	Mod. Poly, Ell	`lat_1, lat_2, lon_1`
kavN	N=5 or 7 for Kavraisky V or VII	Pcyl Sph	–
laea	Lambert Azimuthal Eq. Area	Azi Sph&Ell	–

(continued)

Table C.1 (continued)

Name	Description	Type	Arguments
lonlat	Lon/Lat (geodetic)		–
latlon	Lon/Lat (geodetic alias)		–
lcc	Lambert Conformal Conic	Conic, Sph&Ell	lat_1, lat_2, lat_0
leac	Lambert Equal Area Conic	Conic, Sph&Ell	lat_1, south
loxim	Loximuthal	Pcyl Sph	–

The order and notation follow the PROJ.4 documentation (version 4.9.3), with "Azi" for azimuthal, "Cyl" for cylindrical, "Ell" for elliptical, "Pcyl" for pseudocylindrical, and "Sph" for spherical. Only projections with inverses are included in oce

Table C.2 Projections

Name	Description	Type	Arguments
mbt_s	McBryde-Thomas flat-polar sine (1)	Pcyl, Sph	–
mbt_fps	" " sine (2)	Pcyl, Sph	–
mbtfpp	" " parabolic	Cyl, Sph	–
mbtfpq	" " quartic	Cyl, Sph	–
mbtfps	" " sinusoidal	Pcyl, Sph	–
merc	Mercator	Cyl, Sph&Ell	lat_ts
mil_os	Miller Oblated Stereo.	Azi(mod)	
mill	Miller Cylindrical	Cyl, Sph	
moll	Mollweide	Pcyl, Sph	
murdN	N=1 to 3 for Murdock I to III	Conic, Sph	lat_1, lat_2
natearth	Natural Earth	PCyl, Sph	–
nell	Nell	PCyl, Sph	–
nell_h	Nell-Hammer	PCyl, Sph	–
nsper	Near-sided perspective	Azi, Sph	h
ob_tran	General oblique transformation	Misc Sph	(many)
ocea	Oblique cylindrical equal area	Cyl	lat_1, lat_2, lon_1, lon_2
oea	Oblated Equal Area	Misc Sph	n, m, theta
omerc	Oblique Mercator	Cyl, Sph&Ell	(many)
ortho	Orthographic	Azi, Sph	–
pconic	Perspectictive Conic	Conic, Sph	lat_1, lat_2
poly	Polyconic (American)	Conic, Sph&Ell	
putpN	N=1:5, 3p, 4p, 5p, 6p Putniņš	Pcyl, Sph	–
qua_aut	Quartic Authalic	Pcyl, Sph	–
qsc	Quadrilaterized Spherical cube	Pcyl, Sph	–
robin	Robinson	Pcyl, Sph	–
rouss	Roussilhe Stereographic	Azi, Ell	–
sinu	Sinusoidal (Sanson-Flamsteed)	Pcyl, Sph&Ell	–
somerc	Swiss. Obl. Mercator	Cyl, Ell	–

(continued)

Table C.2 (continued)

Name	Description	Type	Arguments
stere	Stereographic	Azi, Sph&Ell	`lat_ts`
sterea	Oblique Stereographic alternative	Azi, Sph&Ell	–
tcea	Transverse Cyl. equal area	Cyl, Sph	–
tissot	Tissot	Conic, Sph	`lat_1`, `lat_2`
tmerc	Transverse Mercator	Cyl, Sph&Ell	
tpeqd	Two-point equidistant	Misc Sph	`lat_1`, `lat_2`, `lon_1`, `lon_2`
tpers	Tilted perspective	Azi, Sph	`tilt`, `azi`, `h`
ups	Universal polar stereographic	Azi, Sph&Ell	`south`
urmfps	Urmaev flat polar sinusoidal	Pcyl, Sph	`n`
utm	Universal transverse Mercator	Cyl, Sph	`zone`, `south`
vandg	van der Grinten (I)	Misc Sph	
vitk1	Vitkovsky I	Conic, Sph	`lat_1`, `lat_2`
wagN	N=1 to 6 for Wagner I to VI	Pcyl, Sph	`lat_ts` for N=3
weren	Werenskiold I	Pcyl, Sph	
wink1	Winkel I	Pcyl, Sph	`lat_ts`
wintri	Winkel Tripel	Misc Sph	`lat_1`

Appendix D
Seawater Formulations in oce

The oceanographic community has a choice of two sets of formulae for calculating seawater properties: the UNESCO formulation, popularized in the 1980s, and the Gibbs-SeaWater (GSW) formulation, proposed in 2010. This appendix outlines how to navigate between these two systems in the oce package.

In the 1980s, the United Nations Educational, Scientific and Cultural Organization (UNESCO) endorsed a set of formulae for calculating various properties of seawater, including an equation of state that relates density to salinity, temperature and pressure. These formulae are described in most oceanographic textbooks published in the past two decades, and in primary contributions such as those by Fofonoff and Millard (1983) and Millard (1987). As expressed in popular programming languages, the UNESCO formulae were adopted widely by the oceanographic community by the early 1990s.

Recently, a new set of formulae has been made available, known as the Thermodynamic Equation Of Seawater (TEOS-10, where the numbers indicate the year of formulation). The scientific and practical foundations for this are described by IOC et al. (2010), Wright et al. (2011), McDougall and Barker (2011), Graham and McDougall (2012), while Millero (2010) and Pawlowicz et al. (2012) provide historical context. Refinements are expected as new data are acquired, e.g. Budéus (2018) reported recently that new density measurements suggest a possible density bias of order $0.010 \, \text{kg/m}^3$, subject to confirmation by additional studies.

The seawater component of TEOS-10 is denoted GSW, a sort of acronym for Gibbs SeaWater. This work addresses two requirements that arise from new research and improvements in ocean measurement. The first is to put the interrelationships between quantities on a solid thermodynamical basis through the use of a Gibbs function, and the second is to account for spatial variation in the ionic composition of seawater.

In the GSW system, salinity is described in several different ways, of which the most important for oceanographers is a quantity called Absolute Salinity, denoted S_A, that depends on measured water properties *and* the position of measurement.

© Springer Science+Business Media, LLC, part of Springer Nature 2018
D. E. Kelley, *Oceanographic Analysis with R*,
https://doi.org/10.1007/978-1-4939-8844-0

The position is needed to account for spatial variations in seawater ion ratios, assumed invariant of time for the intended range of application. The formulation also proposes a quantity called Conservative Temperature, denoted Θ, which is based on a careful consideration of thermodynamics.

To ease the transition from UNESCO to GSW, the latter has been made available in C, Fortran, Matlab and R. The R case is handled by the gsw package (Kelley et al. 2017), which is loaded automatically by oce.

A simple example may help to solidify the ideas. Suppose a CTD is lowered at 300°E and 30°N (in the central North Atlantic) and it records practical salinity 35 at pressure 1000 dbar (roughly 1 km depth). Then the Absolute Salinity may be calculated with gsw_SA_from_SP() from the gsw package,

```
library(gsw)
SA <- gsw_SA_from_SP(SP=35,p=1000,longitude=300,
latitude=30)
```

yielding S_A=35.1682 g/kg. Similarly, if the in situ temperature were 10°C, the Conservative Temperature could be calculated with[1]

```
CT <- gsw_CT_from_t(SA=SA, t=10, p=1000)
```

yielding Θ=9.869°C.

Conservative Temperature, Θ, should not be confused with potential temperature, θ. To illustrate, the potential temperature referenced to the surface is given by a direct call to a gsw function

```
gsw_pt_from_t(SA=SA, t=10, p=1000, p_ref=0)
| [1] 9.879145
```

or, equivalently, to an oce function (note that position must be specified if eos is "gsw", but not if it is "unesco")

```
swTheta(salinity=35, temperature=10, pressure=1000,
          longitude=300, latitude=30, eos="gsw")
| [1] 9.879145
```

However, swTheta() and gsw_pt_from_t() yield identical results because the former calls the latter.

It should also be noted that the TEOS-10 formulation of potential temperature differs slightly from the UNESCO value

```
swTheta(salinity=35,temperature=10,pressure=1000,
eos="unesco")
| [1] 9.879276
```

although this may be small in many applications.

These examples should be enough to acquaint readers with the gist of using GSW in some common tasks. Much of that work will probably be done with the oce package, which loads gsw on startup. Most oce functions that provide

---

[1]In situ temperature is denoted t in the gsw functions.

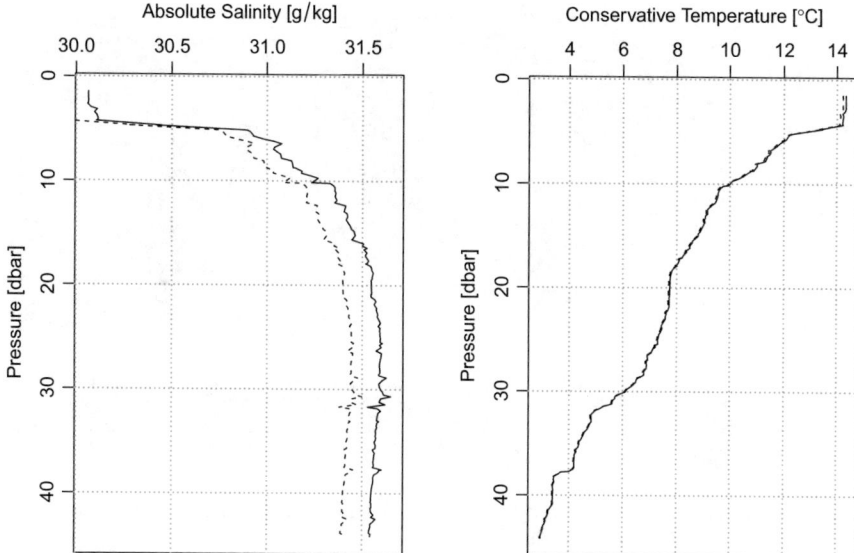

Fig. D.1 Comparison of GSW and UNESCO formulae for seawater properties, using temperature and salinity profiles of the ctd dataset provided by the oce packages. Left: GSW Absolute Salinity (solid) and UNESCO practical salinity (dashed). Right: GSW Conservative Temperature (solid) and UNESCO potential temperature (dashed). Note that the temperature traces are difficult to distinguish at this scale, but the salinity traces differ by much more than the scatter in the measurements

GSW functionality also provide UNESCO functionality. A default preference for UNESCO or GSW formulation may be specified with

```
options(oceEOS="unesco")
```
or
```
options(oceEOS="gsw")
```
which may be done in a startup file (Sect. 2.2.4).

Figure D.1 may help to put the differences between the UNESCO and GSW systems into context. To create the diagram, start with the ctd dataset

```
data(ctd, package="oce")
```
and plot salinity on the left, with Absolute Salinity S_A as a solid line

```
plotProfile(ctd, "salinity", eos="gsw")
```
and practical salinity as a dashed line.

```
lines(ctd[["salinity"]], ctd[["pressure"]],
lty="dashed")
```
Similar code created the temperature panel of Fig. D.1, again with solid and dashed lines for the GSW and UNESCO formulations.

For this particular CTD profile, the differences in the two temperature formulations are small compared with the total variation over depth. By contrast, the salinity changes account for an appreciable fraction of the range over depth, even though

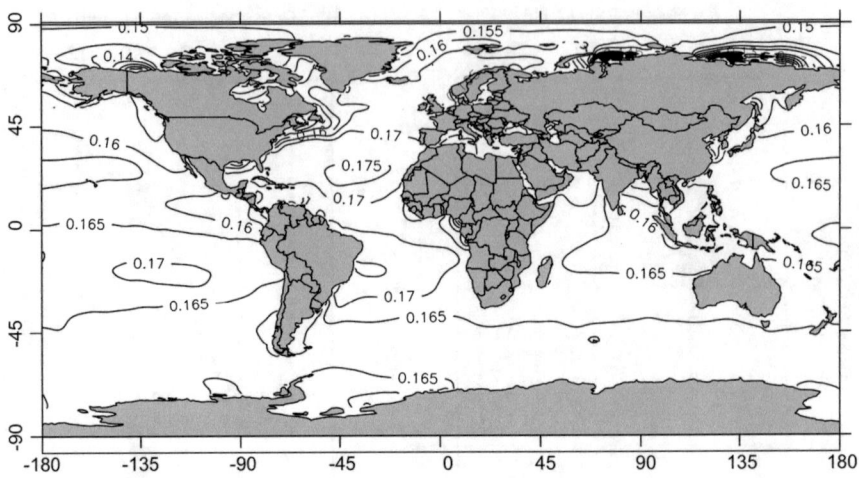

Fig. D.2 Excess of Absolute Salinity over practical salinity at the surface, for the levitus dataset provided by the ocedata package

this sample was taken near a river. Clearly, today's oceanographers must be careful in reporting water properties, to avoid confusion in comparing with observations reported in the older literature, in which the UNESCO system was so engrained.

One way to gain a wider appreciation for the practical difference between UNESCO and GSW is to construct maps, such as that in Fig. D.2, which is based on the levitus dataset, as follows.

```
data(levitus, package="ocedata")
SP <- levitus$SSS
lon <- levitus$longitude
lat <- levitus$latitude
ll <- expand.grid(lon=lon, lat=lat)
SA <- gsw_SA_from_SP(SP=SP, p=0,
                     longitude=ll$lon, latitude=ll$lat)
data(coastlineWorld, package="oce")
plot(coastlineWorld)
contour(lon, lat, SA - SP, labcex=0.8, add=TRUE,
nlevels=30)
```

The resultant diagram shows that the difference between surface Absolute Salinity and practical salinity varies somewhat through space. The pattern is somewhat reminiscent of surface salinity itself, suggesting the two quantities to be in nearly constant ratio. Indeed,

```
summary(as.vector(SA / SP), na.rm=TRUE)
   Min. 1st Qu.  Median    Mean 3rd Qu.    Max.     NA's
  1.005   1.005   1.005   1.005   1.005   1.019   23712
```

shows that the ratio has a limited span for the surface waters described in the levitus dataset.

Appendix E
High-Performance Calculations

Most interpreted languages have limitations when it comes to high-performance computation, and R is no exception. There are issues involving both computation time and memory use. Strategies for diagnosing such issues are sketched in this appendix, along with a few practical methods for overcoming them (Fig. E.1).

Efficiency Limitations of High-Level Languages

R insulates users from the machine in ways that may surprise those who are more accustomed to compiled languages such as Fortran, C and C++. This insulation can be both a blessing and a curse. It saves time that would otherwise be spent programming low-level operations, such as memory allocation, that are difficult to code and maintain. However, it costs time and memory.

For most everyday problems, analysts will be better served using a high-level language such as R than they would be if they took the time to recode algorithms in faster low-level languages. However, some problems are so simply demanding that their effective solution demands some low-level coding, and this appendix provides examples of how to do this.

In addition to truly demanding problems, there are some problems that could be labelled as artificially demanding. The latter may be encountered by inexperienced users with limited understanding of how computers function. For example, high-level languages tend to make it easy to work with arrays, and this can sometimes lead programmers to express algorithms in array form even when it is not actually required. This can yield elegant solutions that work well during the development process, but that may fail badly when applied to a larger actual application (see e.g. Ihaka 2010).

Although honing for efficiency may be an art best learned through experience, there are some simple steps that may make that experience less disruptive.

© Springer Science+Business Media, LLC, part of Springer Nature 2018
D. E. Kelley, *Oceanographic Analysis with R*,
https://doi.org/10.1007/978-1-4939-8844-0

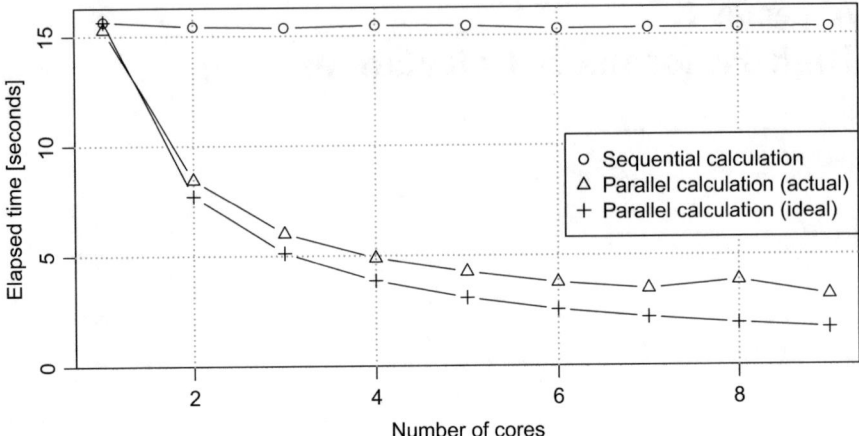

Fig. E.1 Test of multi-core performance with the `doParallel` and `foreach` packages, on a computer with two quad-core processors

Deciding when to optimize

Few programmers would choose an inefficient algorithm if an efficient one is familiar and readily available, but it can be difficult to know at an early stage which algorithms best suit a problem. Furthermore, sometimes it is only when making the transition from exploratory work to routine work that inefficiencies are seen.

Identifying rate-limiting steps is at the core of optimization. For example, speeding a numerical calculation by 10% will not be of great benefit if 99% of the computation time is spent waiting for data transfer. The setting also matters. A commercial enterprise might dedicate a person-month to speed up a task that needlessly takes tens of seconds of user time, but this would be silly in a research setting, where tasks are performed only a few times, and where the user may also be the programmer.

The best decisions on where to direct effort differ from one application to another. Oftentimes, it make senses to start with low-hanging fruit in the R realm, before moving on to lower-level languages.

Avoiding loops

In R, as in many interpreted languages, loops can be significantly slower than vectorized code. The `microbenchmark()` function from the `microbenchmark` package provides an easy way to test the performance differences. Consider the problem of computing $\sum_i i^2$, which can be written as a loop

```
a <- function(i) {
    rval <- 0
    for (ii in i)
        rval <- rval + ii^2
    rval
}
```

or in vectorized form

```
b <- function(i)
    sum(i^2)
```

Using `microbenchmark()` reveals that the looping form of a() is an order of magnitude slower than the vectorized form of b(). However, the squaring operation is quick, and the two methods would have more nearly equal performance for slower operations.

Sometimes, converting loops to vectorized schemes is sensible and efficient, but other times it reduces both code clarity and performance (see e.g. Ihaka 2010). R offers several ways to avoid loops, with the `apply()` family of functions in the R base system, and related functions in the `plyr` package.

R tends gets to get more efficient over time, and it is useful to track how this is done. For example, in version 3.4.0, R altered the memory allocation scheme for assignments made past the end of a vector. Previously, space had been set aside to just satisfy the request, but version 3.4.0 set aside extra memory, in case it might be needed. Thus,

```
A <- function(n) {
    res <- NULL
    for (i in 1:n)
        res[i] <- i^2
    res
}
```

would trigger this extra allocation, whereas

```
B <- function(n) {
    res <- NULL
    for (i in 1:n)
        res <- c(res, i^2)
    res
}
```

would not. As of R version 3.4.0, the A() method is several hundred times faster than the B() method, on the author's computer. This test suggests dropping the use of c() for creating expanding vectors, which might otherwise seem to be a more elegant approach.

Supplementing R with compiled languages

R has interfaces that permit connection with compiled languages such as C and Fortran, and using them can greatly reduce programmer effort and execution time.

There are two basic schemes for connecting R with compiled code (see the R documentation[1] for details). The first uses largely unmodified C or C++ code. For example, the following computes an alternating sum $\sum_1^n (-1)^n x_i$ in C, where x is a vector of numerical values:

```c
void altsum1(int *n, double *x, double *value)
{
    int sign = -1;
    *value = 0.0;
    for (int i = 0; i < *n; i++) {
        *value += sign * x[i];
        sign *= -1;
    }
}
```

Note that the computed value is stored in the `value` argument, not in a returned value. If this code is in a file named `altsum1.c`, then it can be built as a shared library and loaded into R with

```r
system("R CMD SHLIB altsum1.c")
dyn.load("altsum1.so")
```

after which it can be called with, e.g.,

```r
x <- 1:10
.C("altsum1", n=as.integer(length(x)), x=as.double(x),
    value=double(1))$value
 [1]  5
```

The first argument to `.C()` is the name of the function, and the others are the function arguments, which are put in the requisite call-by-reference C form with `as.integer()` and `as.double()`. Space is set aside for the calculated value with `double()`. Since `.C()` returns a list of the arguments supplied to C, the computed result is accessed with the dollar operator.

In the second scheme, C or C++ works directly with R objects. There are two popular approaches to this. In the older one, still used throughout R itself, C macros are used to work with R objects, e.g.

```c
#include <R.h>
#include <Rdefines.h>
SEXP altsum2(SEXP x)
{
    int n= length(x);
    PROTECT(x = AS_NUMERIC(x));
```

---

[1] https://cran.r-project.org/doc/manuals/r-release/R-exts.pdf.

```
      double *xp = REAL(x);
      SEXP rval;
      PROTECT(rval = NEW_NUMERIC(1));
      double *value = NUMERIC_POINTER(rval);
      int sign = -1;
      *value = 0.0;
      for (int i = 0; i < n; i++) {
          *value += sign * xp[i];
          sign *= -1;
      }
      UNPROTECT(2);
      return(rval);
}
```

where the uppercase words are macros that connect C with R. After similar compilation, this is called with a different interface:

```
.Call("altsum2", x)
[1] 5
```

In this simple example, the macros for the R-C++ interface are a significant distraction from the actual algorithm. Although this distraction diminishes with program size, it is not trivial in any case. Luckily, there is another way to connect R with C++, using the Rcpp system (Eddelbuettel and Francois 2011; Eddelbuettel 2013; Wickham 2014).

Using Rcpp, the present example becomes

```
#include <Rcpp.h>
using namespace Rcpp;
// [[Rcpp::export]]
double altsum3(NumericVector x)
{
  double value = 0;
  int sign = -1;
  for (int i = 0; i < x.size(); i++) {
    value += sign * x[i];
    sign *= -1;
  }
  return(value);
}
```

This may be compiled within R with[2]

```
Rcpp::sourceCpp("altsum3.cpp")
```

and run (in R) with

```
altsum3(x)
[1] 5
```

---

[2]The related cppFunction() works with text strings instead of source files, which is convenient for self-contained work.

The simplicity of this example may be enough to illustrate why Rcpp is the first choice of many developers. The related RcppArmadillo system offers further advantages for matrix-heavy work (Eddelbuettel and Sanderson 2014).

A speedup by two orders of magnitude is typical when compiled languages are used instead of R. Even so, this is a method best employed by those who are comfortable with low-level languages, and familiar with dealing with the application crashes that can result when C++ code fails. As noted previously, it is best to stick with R unless there is a real need for the speed and memory advantages offered by C++.

Dealing with memory limitations

R stores objects in random access memory. This causes problems for large objects, particularly because many operations in R involve copying objects before doing calculations with them. For example, R uses a call by value model for the arguments to functions, which means that arguments are duplicated when used within a function. The bigmemory package addresses this by emulating a call by reference model, which may alleviate the problem in some applications. Another scheme is to avoid creating objects that are large, e.g. by only working with selected portions of data stored in a database or other file.

The analysis of algorithms

It is disheartening to spend time writing code to carry out an algorithm that works well on test data, but that proves to be unacceptably slow for the real application of interest. In cases where a variety of algorithms could be employed, it is wise to spend time on the analysis of algorithms before starting to write code. The procedure should be continued during the coding phase, to verify that the analysis was correct. (This latter point is especially important for readers who are coming to R from compiled languages, who may be surprised at the amount of time R can spend copying data, e.g. when calling functions directly or indirectly. It can take a while to learn just which parts of an algorithm will be expensive in R.)

Focus should be directed to how an algorithm performs with large data sets, long integrations, etc., because there is no point speeding up operations that are already fast enough to be practical. Using the symbol n to indicate the size of the problem being tackled, the interest is in predicting the asymptotic resource requirements as n increases. This is similar to the study of the asymptotic behaviour of a function in mathematics.

There are several approaches to this analysis in computing, expressed in Θ notation, $\mathcal{O}$ notation, o notation, and Ω notation (Knuth 1976, 1968). For the present purpose, the most useful of them may be the Θ notation.

Fig. E.2 Definition sketch
for Θ notation, showing that
$T(n)$ falls within bounds
$Cf(n)$ and $C'f(n)$ for $n > n_0$

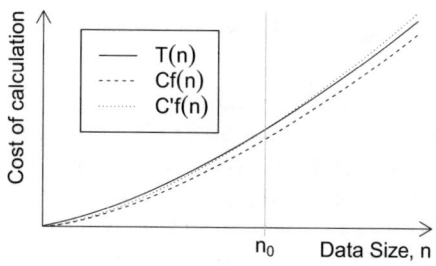

The basic ideas of Θ notation are illustrated in Fig. E.2. Suppose that a computing task has a resource cost T (perhaps in terms of time or memory requirements) that is related to the problem size n by some function $T = T(n)$. For practical purposes, it makes sense to restrict attention to large values of n, say $n > n_0$ say, where n_0 is a constant. Then, it is said that $T(n) = \Theta(f(n))$ if $T(n)$ is bounded between $Cf(n)$ and $C'f(n)$, where C and C' are constants and n is large. Roughly speaking, this means that $T(n)$ is proportional to $f(n)$ for large values of n. The values of C and C' relate to the speed of the computer, etc., and are not of interest in comparing algorithms. For sufficiently large problems, there is also no interest in the value of n_0. The key to the analysis is to focus on the form of $f(n)$.

For example, suppose that $T = 0.01n^{3/2} + n^{1/2}$. Clearly, the second term will be larger than the first if n is small, but the reverse holds for large n. Eventually, as n increases, the first term will overwhelm the second, so that T will approach $0.01n^{3/2}$ as n increases. The notation $T = \Theta(n^{3/2})$ states this. (Note that the scale factor is dropped, which is sensible because the goal is not to predict running time, but rather to indicate the pattern of its increase as a function of problem size, n.)

An algorithm that requires mere examination of data is likely to be $\Theta(n)$, where n is a measure of the size of the dataset. By contrast, an algorithm that involves inter-comparing all the data elements may be $\Theta(n^2)$, which is *much* worse for large datasets. For example, finding the smallest element of a vector is $\Theta(n)$. Sorting such a vector by the bubble sort algorithm is $\Theta(n^2)$, because it involves n steps in which successive minima are found. This makes bubble sort an impractical solution for large data sets, which explains why R provides two other methods: shell sort, which is $\Theta(n \log^2 n)$, and quick sort, which is $\Theta(n \log n)$. These different formulae are not just of academic interest, because modern oceanographic instruments may record several gigabytes in a single deployment, so a $\Theta(n \log n)$ method might be dramatically faster than a $\Theta(n^2)$ algorithm.

It makes sense to consider performance also in terms of memory usage, to avoid the discouragement that might come from spending time developing an algorithm that performs well on small test cases but that overwhelms system memory when applied to actual data. Using `Rprofmem()` during development can help to avoid such problems.

Multiple-processing calculations

Until recent years, advances in chip design have permitted an exponential increase in CPU clock rates over time. This is no longer the case, so much of the research on high-performance computing now centres on harnessing multiple processors, whether co-located on a chip or in a warehouse full of computers. Each scale poses its own challenges, and this is not the place to propose general solutions. Today, it is common to have more than one CPU in a desktop computer, and two or more cores within each CPU, and the discussion will be limited to this circumstance.

Several R packages can be used for parallel processing, and these seem certain to improve over time. (Consult the CRAN high-performance website[3] for this.) Even at this early stage, R can handle multiple cores reasonably well. As an illustration, a calculation was done with the doParallel and foreach packages, on the author's 8-core desktop computer. As Fig. E.1 shows, the performance in this test was close to being linear in the number of cores used.[4] However, it must be noted that this test case did not provide a challenge to the RAM in the machine; if each of those cores required anywhere near 1/8-th of the computer's RAM, the operating system might start paging memory to disk, incurring a significant drop in performance and revealing the need for a different distributed-computation environment.

---

[3] http://cran.r-project.org/web/views/HighPerformanceComputing.html.

[4] Performance achieved with the doMC package was similar, but doParallel may be preferred because doMC is not presently available on the Microsoft Windows platform.

Appendix F
The Future of R

For nearly two decades, R has enjoyed a reputation as a mature language that evolves safely and systematically, with bug fixes and minor additions being provided in new versions that appear several times per year. With a dedicated core of developers working at an impressive pace (illustrated below), we can expect continued refinements to R in the near term. However, mature software systems that are relied upon by large communities are a bit like large ships: they are hard to turn. This is a concern because, as has been noted in this book and elsewhere, R has some weaknesses compared with competitive systems. The history of the language, the open-source license, and the evident talents of the core development team all suggest that the requisite improvements will be made eventually. The question is whether they will be accomplished by continued incremental changes to R, or by a significant shift in the underlying code or even the syntax of the language.

The R source code is handled by a version-control system, and a record of these changes is made available on the R website (see e.g. Fig. F.1). In addition to official releases, daily builds provide users with up-to-date updates on the development version. These things reveal a software product in vigorous and sustained development, and the fact that new releases seldom introduce serious new bugs is clear evidence of the talents of the developers.

Although the message of this book is that R is well suited for oceanographic analysis, there are still opportunities for improvement.

- The base graphics system is good for publication, but lacks interactive and three-dimensional features that are provided in other graphical systems.
- The existence of competing graphical systems imposes some learning burden on the user. For example, many beginners favour the sophisticated power of the `ggplot2` system, but the core-graphics system is used in most R documentation, so there is a need to become conversant in both systems. (The `oce` package relies on the base-graphics system, which alleviates this problem somewhat.)
- Loops can be so slow in R that analysts may need to use C, C++ or Fortran for heavy tasks. Still, for medium-scale problems, R improves year by year. For

© Springer Science+Business Media, LLC, part of Springer Nature 2018
D. E. Kelley, *Oceanographic Analysis with R*,
https://doi.org/10.1007/978-1-4939-8844-0

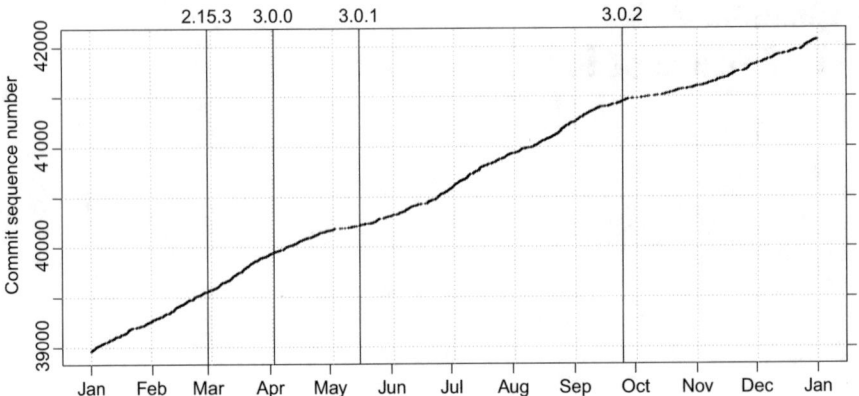

Fig. F.1 Sequence numbers of "commits" to the R source code during the year 2013, counting from zero in January 2003. Vertical lines indicate official R releases

example, the "just in time" compilation and pre-allocation for growing vectors that came in version 3.4.0 yielded noticeable speedups.

- The limitation of data types can be a problem in representing data economically. For example, R lacks the 16-bit integer type that is used by several oceanographic instruments, and promoting these to full machine-length integers can yield significant memory pressure. (This is why the oce package stores some data as pairs of 8-bit sequences.)
- The system of copying data passed to functions has advantages in removing "side effects" but it may sometimes lead to slow performance and excessive memory use. (The memory requirement is subtle, and it depends on how the data are used in the function; R uses clever schemes to avoid bloat.)
- The existence of several models for object orientation in R can be confusing to those learning how to program in the language.

This list of issues is not unique to oceanography, nor restricted to the experience of the present author. Indeed, the list partly derives from insights presented in essays and lectures by Ross Ihaka, one of the original authors of R. Readers interested in the issues and how they may be dealt with in R or a successor language should consult some of Ihaka's publications (Ihaka and Gentleman 1996; Ihaka and Lang 2008; Ihaka 2010), along with other materials provided on Ihaka's website.[1]

---

[1] http://www.stat.auckland.ac.nz/~ihaka/.

References

Airy, G. B., 1861. Explanation of a projection by balance of errors for maps applying to a very large extent of the earth's surface; and comparison of this projection with other projections. *London, Edinburgh and Dublin Philosophical Magazine and Journal of Science (4th ser.)*, 22(149):409–421.

Albert, J., 2009. *Bayesian computation with R*. Use R! Springer, New York, NY, USA, second edition.

Anderson, T. R. and Gentleman, W. C., 2012. The legacy of Gordon Arthur Riley (1911–1985) and the development of mathematical models in biological oceanography. *Journal of Marine Research*, 70(1):1–30.

Antonov, J. I., Seidov, D., Bower, T. P., Locarnini, R. A., Mishonov, A. V., Garcia, H. E., Baranova, O. K., Zweng, M. M., and Johnson, D. R., 2010. World ocean atlas 2009. Technical report, US Government printing Office.

Ashmead, J., 2012. Morlet wavelets in quantum mechanics. *Quanta*, 1(1):58–70.

Bååth, R., 2012. The state of naming conventions in R. *The R Journal*, 4(2):74–75.

Backus, G. E., 1964. Magnetic anomalies over oceanic ridges. *Nature*, 4919:591–592.

Barnes, S. L., 1994. Application of Barnes objective analysis scheme. Part I: effects of undersampling, wave position, and station randomness. *Journal of Atmospheric and Oceanic Technology*, 11:1433–1447.

Barnett, T. P. and Hasselmann, K., 1979. Techniques of linear prediction, with application to oceanic and atmospheric fields in the tropical Pacific. *Reviews of Geophysics*, 17(5):949–968.

Barsi, J. A., Schott, J. R., Palluconi, F. D., Helder, D. L., Hook, S. J., Markham, B. L., Chander, G., and O'Donnel, E. M., 2003. Landsat TM and ETM+ thermal band calibration. *Canadian Journal of Remote Sensing*, 29(2):141–153.

Becker, R. A. and Chambers, J. M., 1984. *S: an interactive environment for data analysis and graphics*. Wadsworth statistics/probability series. Wadsworth Advanced Book Program, Belmont, CA, USA.

Becker, R. A., Chambers, J. M., and Wilks, A. R., 1988. *The new S language*. Wadsworth & Brooks/Cole, Pacific Grove, CA, USA.

Bergmeir, C. and Benítez, J., 2012. Neural networks in R using the Stuttgart Neural Network Simulator: RSNNS. *Journal of Statistical Software, Articles*, 46(7):1–26.

Bloomfield, P., 2005. *Fourier analysis of time series: an introduction*. Probability and statistics. Wiley Press, Hoboken, NJ, USA.

Boddy, L., Morris, C., Wilkins, M., Al-Haddad, L., Tarran, G., Jonker, R., and Burkill, P., 2000. Identification of 72 phytoplankon species by radial basis function neural network analysis of flow cytometric data. *Marine Ecology Progress Series*, 195:47–59.

© Springer Science+Business Media, LLC, part of Springer Nature 2018 269
D. E. Kelley, *Oceanographic Analysis with R*,
https://doi.org/10.1007/978-1-4939-8844-0

Boddy, L., Morris, C., Wilkins, M., Tarran, G., and Burkill, P., 1994. Neural network analysis of flow cytometric data for 40 marine phytoplankton species. *Cytometry*, 15:283–293.

Borcard, D., Gillet, F., and Legendre, P., 2011. *Numerical Ecology with R*. Use R. Springer-Verlag, New York, NY, USA.

Bourgault, D., Cyr, F., Dumont, D., and Carter, A., 2014. Numerical simulations of the spread of floating passive tracer released at the Old Harry prospect. *Environmental Research Letters*, 9:054001.

Bourgault, D. and Koutitonsky, V. G., 1999. Real-time monitoring of the freshwater discharge at the head of the St. Lawrence Estuary. *Atmosphere-Ocean*, 37(2):203–220.

Box, G. E. P. and Jenkins, G. M., 1976. *Time series analysis forecasting and control*. Holden-Day, Oakland, CA, USA.

Boyer, T. P., Antonov, J. I., Baranova, O. K., Garcia, H. E., Johnson, D. R., Locarnini, R. A., Mishonov, A. V., O'Brien, T. D., Seidov, D., Smolyar, V., and Zweng, M. M., 2009. World ocean atlas 2009. Technical report, US Government printing Office.

Brainerd, K. E. and Gregg, M. C., 1995. Surface mixed and mixing layer depths. *Deep Sea Research Part I: Oceanographic Research Papers*, 42(9):1521–1543.

Bretherton, C. S., Smith, C., and Wallace, J. M., 1992. An intercomparison of methods for finding coupled patterns in climate data. *Journal of Climate*, 5(6):541–560.

Bretherton, F. P., Davis, R. E., and Fandry, C. B., 1976. A technique for objective analysis and design of oceanographic experiments applied to MODE-73. *Deep-Sea Research*, 23:559–582.

Brillinger, D. R., 1981. *Time series: data analysis and theory*. Holden-Day series in time series analysis. Holden-Day, San Francisco, CA, USA, expanded edition.

Broecker, W. S., 1991. The great ocean conveyor. *Oceanography*, 4(2):79–89.

Budéus, G. T., 2018. Potential bias in TEOS10 density of sea water samples. *Deep-Sea Research Part 1*, 134:41–47.

Cahill, N., Rahmstorf, S., and Parnell, A. C., 2015. Change points of global temperature. *Environmental Research Letters*, 10(8):084002.

Carr, D. B., 1991. Looking at large data sets using binned data plots. In Buja, A. and Tukey, P. A., editors, *Computing and Graphics in Statistics*, pages 7–39. Springer-Verlag New York, Inc., New York, NY, USA.

Carter, E. F. and Robinson, A. R., 1987. Analysis methods for the estimation of oceanic fields. *Journal of Atmospheric and Oceanic Technology*, 4(1):49–74.

Chambers, J. M., 2008. *Software for data analysis: programming with R*. Statistics and computing. Springer-Verlag, New York, NY, USA.

Chambers, J. M. and Hastie, T. J., 1992. *Statistical models in S*. Wadsworth & Brooks/Cole, Pacific Grove, CA, USA.

Chu, P. and Fan, C., 2010a. Objective determination of global ocean surface mixed layer depth. In *Proceedings on MTS/IEEE OCEANS 10*, pages 1001–1007, Seattle, WA, USA.

Chu, P. C. and Fan, C., 2010b. Optimal linear fitting for objective determination of ocean mixed layer depth from glider profiles. *Journal of Atmospheric and Oceanic Technology*, 27(11):1893–1898.

Clarke, A. J. and Van Gorder, S., 2012. On fitting a straight line to data when the "noise" in both variables is unknown. *Journal of Atmospheric and Oceanic Technology*, 30(1):151–158.

Cleveland, W. S. and McGill, R., 1984. Graphical perception: Theory, experimentation, and application to the development of graphical methods. *Journal of the American Statistical Association*, 79(387):531–554.

Culkin, F. and Smith, N. D., 1980. Determination of the concentration of potassium chloride solution having the same electrical conductivity, at 15°C and infinite frequency, as standard seawater of salinity 35.0000‰ (chlorinity 19.37394‰). *IEEE Journal of Oceanic Engineering*, OE-5(1):22–23.

Cushman-Roisin, B. and Beckers, J., 2011. *Introduction to geophysical fluid dynamics*. Academic Press.

Daley, R., 1991. *Atmospheric data analysis*. Cambridge, New York, NY, USA.

Dalgaard, P., 2002. *Introductory Statistics with R.* Statistics and Computing. Springer, New York, NY, USA.

Daubechies, I., 1990. The wavelet transform, time-frequency localization and signal analysis. *IEEE Transactions on Information Theory*, 36(5):961–1005.

Davis, R. E., 1985. Objective mapping by least squares fitting. *Journal of Geophysical Research*, 90(C3):4773–4777.

de Boyer Montégut, C., Madec, G., Fischer, A. S., and Lazar, A., 2004. Mixed layer depth over the global ocean: an examination of profile and profile-based climatology. *Journal of Geophysical Research*, 109(C12003).

De Veaux, R. D., Velleman, P. R., and Bock, D. E., 2006. *Intro Stats.* Pearson Addison Wesley, Boston, MA, USA, 2nd edition.

deYoung, B., Barange, M., Beaugrand, G., Harris, R., Perry, R. I., Scheffer, M., and Werner, F., 2008. Regime shifts in marine ecosystems: detection, prediction and management. *Trends in Ecology & Evolution*, 23(7):402–409.

Drinkwater, K. F., Myers, R. A., Pettipas, R. G., and Wright, T. L., 1994. Climatic data for the northwest Atlantic: the position of the shelf/slope front and the northern boundary of the Gulf Stream between 50w and 75w, 1973–1992. Technical report, Department of Fisheries and Oceans, Dartmouth, NS, Canada.

Dunn, J. R. and Ridgway, K. R., 2002. Mapping ocean properties in regions of complex topography. *Deep Sea Research Part I: Oceanographic Research Papers*, 49(3):591–604.

Eddelbuettel, D., 2013. *Seamless R and C++ Integration with Rcpp.* Springer, New York.

Eddelbuettel, D. and Francois, R., 2011. Seamless R and C++ integration. *Journal of Stastistical Software*, 40(8).

Eddelbuettel, D. and Sanderson, C., 2014. Rcpparmadillo: Accelerating R with high-performance C++ linear algebra. *Computational Statistics & Data Analysis*, 71:1054–1063.

Efron, B. and Gong, G., 1983. A leisurely look at the bootstrap, the jacknife, and cross-validation. *American Statistician*, 37(1):36–48.

Efron, B. and Tibshirani, R. J., 1998. *An introduction to the bootstrap.* On Statistics and Applied Probability. Chapman & Hall/CRC, Boca Raton, FL, USA.

Emery, W. J. and Thomson, R. E., 2001. *Data analysis methods in Physical Oceanography.* Elsevier, Amsterdam.

Estivill-Castro, V., 2002. Why so many clustering algorithms—a position paper. *ACM SIGKDD Explorations Newsletter*, 4(1):65–75.

Evenden, G. I., 2003. Cartographic projection procedures for the UNIX environment – a user's manual. Technical Report 90-284, United States Department of the Interior: Geological Survey, Woods Hole MA USA.

Faraway, J. J., 2002. *Practical regression and ANOVA using R.* The Comprehensive R Archive Network (online).

Faraway, J. J., 2005. *Linear models with R.* Texts in statistical science. Chapman & Hall/CRC, Boca Raton, FL, USA.

Fofonoff, N. P. and Millard, R. C., 1983. Algorithms for computation of fundamental properties of seawater. Unesco technical papers in marine science 44, UNESCO.

Foreman, M. G. G., 1977. Manual for tidal heights analysis and prediction. *Pacific Marine Science Report*, 77(10):1–58.

Foreman, M. G. G. and Neufeld, E. T., 1991. Harmonic tidal analysis of long time series. *International hydrographic review*, 68(1):95–108.

Fox, J., 2009. Aspects of the Social Organization and Trajectory of the R Project. *The R Journal*, 1(2):5–13.

Galbraith, P. S. and Kelley, D. E., 1996. Identifying overturns in CTD profiles. *Journal of Atmospheric and Oceanic Technology*, 13:688–702.

Gallant, A. R., 1975. Nonlinear regression. *The American Statistician*, 29(2):pp. 73–81.

Garratt, J. R., 1977. Review of drag coefficients over oceans and continents. *Monthly Weather Review*, 105:915–927.

Gebbie, G. and Huybers, P., 2010. Total matrix intercomparison: A method for determining the geometry of water-mass pathways. *Journal of Physical Oceanography*, 40:1710–1728.

Gentleman, R. and Ihaka, R., 2000. Lexical scope and statistical computing. *Journal of Computational and Graphical Statistics*, 9(3):pp. 491–508.

Gill, A. E., 1982. *Atmosphere-ocean Dynamics*. Academic Press, New York, NY, USA.

Glover, D. M., Jenkins, W. J., and Doney, S. C., 2011. *Modeling Methods for Marine Science*. Cambridge University Press, Cambridge, UK.

Godfrey, A. J. R., 2013. Statistical analysis from a blind person's perspective. *The R Journal*, 5(1):73–80.

Godfrey, A. J. R. and Erhardt, R., 2014. Addendum to "statistical software from a blind person's perspective". *The R Journal*, 6(1):182–182.

Godin, G., 1972. *The Analysis of Tides*. University of Toronto Press, Toronto, ON, Canada.

Gonella, J., 1972. A rotary-component method for analysing meteorological and oceanographic vector time series. *Deep-Sea Research*, 19:833–846.

Goslee, S. C., 2011. Analyzing remote sensing data in R: the landsat package. *Journal of Statistical Software*, 43(4):1–25.

Graham, F. S. and McDougall, T. J., 2012. Quantifying the nonconservative production of conservative temperature, potential temperature, and entropy. *Journal of Physical Oceanography*, 43(5):838–862.

Grant, H. L., Stewart, R. W., and Moilliet, A., 1962. Turbulence spectra from a tidal channel. *Journal of Fluid Mechanics*, 12(2):241–268.

Gregg, M. C., 1991. The study of mixing in the ocean: a brief history. *Oceanography*, 4(1):39–45.

Grolemund, G. and Wickham, H., 2011. Dates and times made easy with lubridate. *Journal of Statistical Software*, 40(3):1–25.

Gustafsson, F., 1996. Determining the initial states in forward-backward filtering. *IEEE Transactions on Signal Processing*, 44(4):988–992.

Hansen, J., Ruedy, R., Sato, M., and Lo, K., 2010. Global surface temperature change. *Reviews of Geophysics*, 48(4):RG4004.

Harris, F. J., 1978. On the use of windows for harmonic analysis with the discrete Fourier transform. *Proceedings of the IEEE*, 66(1):51–83.

Hartigan, J. A. and Wong, M. A., 1979. A k-means clustering algorithm. *Journal of the Royal Statistical Society. Series C (Applied Statistics)*, 28(1):100–108.

Hauer, E., 2004. The harm done by tests of significance. *Accident Analysis and Prevention*, 36(495–500).

Heiberger, R. M. and Neuwirth, E., 2009. *R through Excel*. Springer, New York, NY, USA.

Helber, R. W., Barron, C. N., Carnes, M. R., and Zingarelli, R. A., 2008. Evaluating the sonic layer depth relative to the mixed layer depth. *Journal of Geophysical Research*, 113(C7).

Herndon, T., Ash, M., and Pollin, R., 2013. Does high public debt consistently stifle economic growth? a critique of Reinhart and Rogoff. Working Paper, 322, Political Economy Research Institute, University of Massachusetts Amherst.

Hinrichsen, H.-H. and Tomczak, M., 1993. Optimum multiparameter analysis of the water mass structure in the Western North Atlantic Ocean. *Journal of Geophysical Research*, 98(C6):10155–10169.

Horton, N. J. and Kleinman, K. P., 2007. Much ado about nothing: a comparison of missing data methods and software to fit incomplete data regression models. *American Statistician*, 61(1):79–90.

Hothorn, T., Hornik, K., and Zeileis, A., 2006. Unbiased recursive partitioning: A conditional inference framework. *Journal of Computational and Graphical Statistics*, 15(3):651–674.

Hsieh, W. W., 2009. *Machine learning methods in the environmental sciences*. Cambridge University Press, Cambridge, UK.

Hsieh, W. W. and Tang, B., 1998. Applying neural network models to prediction and data analysis in meteorology and oceanography. *Bulletin of the American Meteorological Society*, 79(9):1855–1870.

Hutson, M., 2018. Artificial intelligence faces reproducibility crisis. *Science*, 359(6377):725–726.

IEEE Computer Society, 2008. IEEE Standard for Floating-Point Arithmetic. Technical Report 754TM-2008, IEEE, New York NY.

Ihaka, R., 2003. Colour for presentation graphics. In *Proceedings of the 3rd international workshop on distributed statistical computing*, Technische Universität Wien, Vienna, Austria.

Ihaka, R., 2010. R: lessons learned, directions for the future. In *Proceedings of Joint Statistical Meeting 2010*, Vancouver, BC, Canada.

Ihaka, R. and Gentleman, R., 1996. R: A language for data analysis and graphics. *Journal of Computational & Graphical Statistics*, 5(3):pp. 299–314.

Ihaka, R. and Lang, D. T., 2008. Back to the future: Lisp as a base for a statistical computing system. In *Compstat 2008*, Porto, Portugal.

Ingham, M. C., 1966. *The salinity extrema of the world ocean*. PhD thesis, Oregon State University, Corvallis, OR, USA.

Intergovermental Oceanographic Commission, 1985. *Manual on sea level measurement and interpretation, volume 1*. UNESCO, Paris.

IOC, SCOR, and IAPSO, 2010. The international thermodynamic equation of seawater–2010: Calculation and use of thermodynamic properties. Technical Report 56, Intergovernmental Oceanographic Commission, Manuals and Guide.

Jenkins, G. M. and Watts, D. G., 1969. *Spectral analysis and its applications*. Holden-Day, San Francisco, CA, USA.

Jiménez-Muñoz, J. C., Sobrino, J. A., Skoković, D., Mattar, C., and Cristóbal, J., 2014. Land surface temperature retrieval methods from Landsat-8 thermal infrared sensor data. *IEEE Geoscience and Remote Sensing Letters*, 11(10):1840–1843.

Johnson, J. B. and Omland, K. S., 2004. Model selection in ecology and evolution. *Trends in Ecology & Evolution*, 19(2):101–108.

Johnson, N. L., Kotz, S., and Balakrishnan, N., 1995. *Continuous Univariate Distributions*, volume 1. Wiley, New York, NY, USA.

Kara, A. B., Rochford, P. A., and Hurlburt, H. E., 2000. An optimal definition for ocean mixed layer depth. *Journal of Geophysical Research*, 105(C7):16803–16821.

Keeling, C. D., 1960. The concentration and isotopic abundances of carbon dioxide in the atmosphere. *Tellus*, 12(2):200–203.

Kelley, D. and Richards, C., 2018. *oce: Analysis of Oceanographic data*. Comprehensive R Archive Network.

Kelley, D., Richards, C., and SCOR/IAPSO, W., 2017. *gsw: Gibbs Sea Water Functions*.

Kelley, D. E. and Van Scoy, K. A., 1999. A basinwide estimate of vertical mixing in the upper pycnocline: Spreading of bomb tritium in the North Pacific Ocean. *Journal of Physical Oceanography*, 29(8):1759–1771.

Killick, R. and Eckley, I. A., 2014. changepoint: An R package for changepoint analysis. *Journal of Statistical Software*, 58(3):1–19.

Killick, R., Haynes, K., and Eckley, I. A., 2016. *changepoint: An R package for changepoint analysis*.

Knuth, D. E., 1968. *The art of computer programming*, volume 1: Fundamental algorithms. Addison-Wesley, Boston, MA, USA.

Knuth, D. E., 1976. Big omicron and big omega and big theta. *Newsletter, Association of Computing Machines Special Interest Group on Algorithms and Computation Theory*, 8(2).

Koch, S. E., DesJardins, M., and Kocin, P. J., 1983. An interactive Barnes objective map analysis scheme for use with satellite and conventional data. *Journal of Climate and Applied Meteorology*, 22:1487–1503.

Lämmel, R., 2008. Google's MapReduce programming model–revisited. *Science of Computer Programming*, 70(1):1–30.

LeCun, Y., Bengio, Y., and Hinton, G., 2015. Deep learning. *Nature*, 521:436–444.

Legendre, P., 2014. *lmodel2: Model II Regression*. Comprehensive R Archive Network.

Legendre, P. and Legendre, L., 1998. *Numerical Ecology*. Developments in environmental modeling 20. Elsevier, Amsterdam, 2nd English edition.

Leisch, F., 2002. Sweave: Dynamic generation of statistical reports using literate data analysis. In Härdle, W. and Rönz, B., editors, *Compstat 2002 — Proceedings in Computational Statistics*, pages 575–580. Physica Verlag, Heidelberg. ISBN 3-7908-1517-9.

Levitus, S., 1982. Climatological atlas of the world ocean. Technical report, U.S. Govt. printing office, NOAA Prof. paper 13, Washington, DC, USA.

Levitus, S. and Boyer, T., 1994. World ocean atlas 1994 volume 4: Temperature. Technical report, U.S. Department of Commerce, NOAA Atlas NESDIS 4, Washington, DC, USA.

Light, A. and Bartlein, P. J., 2004. The end of the rainbow? Color schemes for improved data graphics. *Eos Trans. AGU*, 85(40).

Limpert, E., Stahel, W. A., and Abbt, M., 2001. Log-normal distributions across the science: keys and clues. *BioScience*, 51(5):341–352.

Lindegren, M., Dakos, V., Gröger, J. P., Gårdmark, A., Kornilovs, G., Otto, S. A., and Möllmann, C., 2012. Early detection of ecosystem regime shifts: A multiple method evaluation for management application. *PLoS ONE*, 7(7):e38410.

Locarnini, R. A., Mishonov, A. V., Antonov, J. I., Boyer, T. P., Garcia, H. E., Baranova, O. K., Zweng, M. M., and Johnson, D. R., 2010. World ocean atlas 2009. Technical report, US Government printing Office.

Lowndes, J. S. S., Best, B. D., Scarborough, C., Afflerbach, J. C., Frazier, M. R., O'Hara, C. C., Jiang, N., and Halpern, B. S., 2017. Our path to better science in less time using open data science tools. *Nature Ecology & Evolution*, 1:0160.

Mallows, C., 2006. Tukey's paper after 40 years. *Technometrics*, 48(3):319–325.

Mamayev, O., Dooley, H., Millard, B., and Taira, K., 1991. *Processing of oceanographic station data*. UNESCO, Paris, France.

Mann, M. E., 2008. Smoothing of climate time series revisited. *Geophysical Research Letters*, 35, L16708.

Marsden, R. F., 1999. A proposal for a neutral regression. *Journal of Atmospheric and Oceanic Technology*, 16(7):876–883.

McArdle, B. H., 2003. Lines, models, and errors: regression in the field. *Limnology and Oceanography*, 48(3):1363–1366.

McDougall, T. J. and Barker, P. M., 2011. *Getting started with TEOS-10 and the Gibbs Seawater (GSW) Oceanographic Toolbox*. SCOR/IAPSO WG127.

McDougall, T. J. and Jackett, D. R., 2007. The thinness of the ocean in S-θ-p space and the implications for mean diapycnal advection. *Journal of Physical Oceanography*, 37(6):1714–1732.

Melet, A., Legg, S., and Hallberg, R., 2016. Climatic impacts of parameterized local and remote tidal mixing. *Journal of Climate*, 29(10):3473–3500.

Millard, R. C., 1987. International oceanographic tables. Unesco technical papers in marine science 40, Unesco.

Miller, A. J., Cayan, D. R., Barnett, T. P., Graham, N. E., and Oberhuber, J. M., 1994. The 1976–77 climate shift of the Pacific Ocean. *Oceanography*, 7(1):21–26.

Millero, F. J., 2010. History of the equation of state of seawater. *Oceanography*, 23(3):18–33.

Mills, E. L., 2009. *The Fluid Envelope of Our Planet: How the Study of Ocean Currents Became a Science*. University of Toronto Press.

Milne, A., 1867. Extract of a letter from Vice-Admiral Sir Alexander Milne, K.C.B., to the President, concerning the currents on the N.E. coast of America. *Proceedings and Transactions of the Nova Scotian Institute of Science*, 1(3):140.

Missan, S., Luo, J., and Trappenberg, T., 2018. Using convolutional neural networks for classification of marine plankton images recorded with a submersible digital holographic microscope. (2018 Ocean Sciences meeting, Portland OR, presentation OD53A-06).

Moritz, S. and Bartz-Beielstein, T., 2017. imputeTS: Time Series Missing Value Imputation in R. *The R Journal*, 9(1):207–218.

Morlet, J., Arens, G., Fourgeau, E., and Giard, D., 1982. Wave propagation and sampling theory-part II: sampling theory and complex waves. *Geophysics*, 47(2):222–236.

Muggeo, V. M. R., 2008. segmented: An R package to fit regression models with broken-line relationships. *R News*, 8(1):20–25.

Munk, W. and Wunsch, C., 1998. Abyssal recipes II. energetics of the tides and wind. *Deep-Sea Research*, 45:1976–2009.

Munk, W. H., 1966. Abyssal recipes. *Deep-Sea Research*, 13:707–730.

Munk, W. H. and Cartwright, D. E., 1966. Tidal spectroscopy and prediction. *Philosophical Transactions of the Royal Society of London. Series A, Mathematical and Physical Sciences*, 259(1105):533–581.

Murrell, P., 2006. *R Graphics*. Chapman & Hall/CRC, Boca Raton, FL, USA.

Nocedal, J. and Wright, S. J., 1999. *Numerical optimization*. Springer series in operations research. Springer-Verlag, New York, NY, USA.

North, G. R., 1984. Empirical orthogonal functions and normal modes. *Journal of the Atmospheric Sciences*, 41(5):879–887.

North, G. R., Bell, T. L., Cahalan, R. F., and Joeng, F. J., 1982. Sampling errors in the estimation of Empirical Orthogonal Functions. *Monthly Weather Review*, 110:699–706.

Osborn, T. R., 1980. Estimates of the local rate of vertical diffusion from dissipation measurements. *Journal of Physical Oceanography*, 10:83–89.

Östlund, H. G. and Hut, G., 1984. Arctic ocean water mass balance from isotope data. *Journal of Geophysical Research: Oceans*, 89(C4):6373–6381.

Pawlowicz, R., Beardsley, B., and Lentz, S., 2002. Classical tidal harmonic analysis including error estimates in Matlab using T_TIDE. *Computers & Geosciences*, 28(8):929–937.

Pawlowicz, R., McDougall, T., and Tailleux, R. I., 2012. An historical perspective on the development of the thermodynamic equation of seawater-2010. *Ocean Science*, 8:161–174.

Pebesma, E., Nüst, D., and Bivand, R., 2012. The R software environment in reproducible geoscientific research. *Eos, Transactions, American Geophysical Union*, 93(16).

Philander, S. G. and Fedorov, A., 2003. Is El Niño sporadic or cyclic? *Annual Reviews of Earth and Planetary Science*, 31:579–594.

Priestley, M. B., 1981. *Spectral analysis and time series*. Academic Press, New York, NY, USA.

Pugh, D. T., 1987. *Tides, surges and mean sea level*. John Wiley and Sons, Chichester, UK.

R Core Team, 2017. *R: A Language and Environment for Statistical Computing*. R Foundation for Statistical Computing, Vienna, Austria.

Raymond, E. S., 2001. *The cathedral and the bazaar: musings on linux and open source by an accidental revolutionary*. O'Reilly and Associates, Sebastopol, CA, USA.

Redfield, A. C., 1934. *James Johnstone Memorial Volume*, On the proportions of organic derivations in sea water and their relation to the composition of plankton, pages 177–192. University Press of Liverpool, Liverpool, UK.

Reiniger, R. F. and Ross, C. K., 1968. A method of interpolation with application to oceanographic data. *Deep-Sea Research*, 15:185–193.

Revelle, R., 1995. Alfred C. Redfield: a biographical memoir. *Biographical Memoirs*, 67:313–329.

Richards, C., Bourgault, D., Galbraith, P. S., Hay, A., and Kelley, D. E., 2013. Measurements of shoaling internal waves and turbulence in an estuary. *Journal of Geophysical Research: Oceans*, 118(1):273–286.

Ricker, W. E., 1973. Linear regressions in fishery research. *Journal of the Fisheries Research Board of Canada*, 30:409–434.

Riley, G. A., 1946. Factors controlling phytoplankton populations on Georges Bank. *Journal of Marine Research*, 6(1):54–73.

Ripley, B. D., 1994. Neural networks and related methods for classification. *Journal of the Royal Statistical Society. Series B (Methodological)*, 56(3):409–456.

Ripley, B. D., 1996. *Pattern recognition and neural networks*. Cambridge University Press, Cambridge, UK.

Ripley, B. D. and Hornik, K., 2001. Date-time classes. *R News*, 1(2):8–11.

Roberts, J. and Roberts, T. D., 1978. Use of the Butterworth low-pass filter for oceanographic data. *Journal of Geophysical Research: Oceans*, 83(C11):5510–5514.

Roesch, A. and Schmidbauer, H., 2014. *WaveletComp: Computational Wavelet Analysis.*

Rudnick, D. L. and Davis, R. E., 2003. Red noise and regime shifts. *Deep Sea Research Part I: Oceanographic Research Papers*, 50(6):691–699.

Sambridge, M., Braun, J., and McQueen, H., 1995. Geophysical parameterization and interpolation of irregular data using natural neighbors. *Geophysical Journal International*, 122(3):837–857.

Sarmiento, J. L. and Gruber, N., 2006. *Ocean Biogeochemical Dynamics*. Princeton University Press, Princeton NJ.

Schevill, W. E. and Watkins, W. A., 1962. Whale and porpoise voices: a phonograph record. Technical report, Woods Hole Oceanographic Institution (contribution 1320).

Schmidtko, S., Johnson, G. C., and Lyman, J. M., 2013. Mimoc: A global monthly isopycnal upper-ocean climatology with mixed layers. *Journal of Geophysical Research: Oceans*, 118(4):1658–1672.

Schmitt, R. W., 1981. Form of the temperature-salinity relationship in central water: evidence for double-diffusive mixing. *Journal of Physical Oceanography*, 11:1015–1026.

Sea-Bird Electronics, 2016. *Seasoft V2: SBE data processing*. Sea-Bird Electronics, Bellevue WA USA, 7.26.5 edition.

Shumway, R. H. and Stoffer, D. S., 2006. *Time Series Analysis and its Applications: With R Examples*. Springer-Verlag, New York, 2nd edition.

Snodgrass, F. E., Groves, G. W., Hasslemann, K. F., Miller, G. R., Munk, W. H., and Powers, W. H., 1966. Propagation of ocean swell across the Pacific. *Phil. Trans. R. Soc. Lond. A*, 259(1103):431–497.

Snyder, J. P., 1987. *Map projections—a working manual*. U.S. Geological survey professional paper 1395, Washington.

Snyder, J. P., 1993. *Two thousand years of map projections*. University of Chicago Press, Chicago, IL.

Sobel, D., 1995. *Longitude*. Walker and Company, New York.

Soetaert, K. and Herman, P. M. J., 2009. *A Practical Guide to Ecological Modelling: Using R as a Simulation Platform*. Springer.

Soetaert, K., Petzoldt, T., and Setzer, R. W., 2010. Solving differential equations in R. *The R Journal*, 2(2):5–15.

Stommel, H., 1948. The westward intensification of wind-driven ocean circulation. *Transactions, American Geophysical Union*, 29(2):202–206.

Stommel, H., 1961. Thermohaline convection with two stable regimes of flow. *Tellus*, 13(2):224–230.

Sverdrup, H. U., 1953. On conditions for the vernal blooming of phytoplankton. *ICES Journal of Marine Science*, 18(3):287–295.

Sverdrup, H. U., Johnson, M. W., and Fleming, R. H., 1942. *The Oceans, Their Physics, Chemistry, and General Biology*. Prentice-Hall, New York.

Talley, L. D., Pickard, G. L., Emery, W. J., and Swift, J. H., 2011. *Descriptive Physical Oceanography*. Elsevier, Amsterdam, sixth edition.

Tangang, F. T., Tang, B., Monahan, A. H., and Hsieh, W. W., 1998. Forecasting ENSO events: A neural network–extended EOF approach. *Journal of Climate*, 11(1):29–41.

Taylor, B. N. and Kuyatt, C. E., 1994. Guidelines for evaluating and expressing the uncertainty of NIST measurement results. NIST Technical Note 1297, U.S. Department of Commerce Technology Administration: National Institute of Standards and Technology, Gaithersburg, MD, USA.

Teledyne-RDI, 2007. *Workhorse commands and output data format*. Teledyne-RDI.

Thomson, R. E. and Fine, I. V., 2003. Estimating mixed layer depth from oceanic profile data. *Journal of Atmospheric and Oceanic Technology*, 20(2):319–329.

Tissot, A., 1881. *Mémoire sur la représentation des surfaces et les projections des cartes géographiques*. Gauthier-Villars, Paris, France.

Tomczak, M., 1999. Some historical, theoretical and applied aspects of quantitative water mass analysis. *Journal of Marine Research*, 57:275–303.

Torrence, C. and Compo, G. P., 1998. A practical guide to wavelet analysis. *Bulletin of the American Meteorological Society*, 79(1):61–78.

Tukey, J. W., 1962. The future of data analysis. *The Annals of Mathematical Statistics*, 32(1):1–67.

Tukey, J. W., 1977. *Exploratory Data Analysis*. Addison-Wesley, Reading, MA, USA.

Turner, J. S., 1973. *Buoyancy Effects in Fluids*. Cambridge University Press, Cambridge, UK.

UNESCO, 1988. The acquisition, calibration and analysis of CTD data. *UNESCO tech. papers in mar. sci.*, 54:92pp.

Velleman, P. F. and Hoaglin, D. C., 2004. *Applications, basics, and computing of Exploratory Data Analysis*. The Internet-First University Press, Ithaca, NY, USA.

Venables, W. N. and Ripley, B. D., 1999. *Modern applied statistics with S-plus*. Springer-Verlag, New York, NY, USA, third edition.

Vine, F. J. and Matthews, D. H., 1963. Magnetic anomalies over oceanic ridges. *Nature*, 4897:947–949.

Wallace, J. M. and Dickinson, R. E., 1972. Empirical orthogonal representation of time series in the frequency domain. Part I: Theoretical considerations. *Journal of Applied Meteorology*, 11:887–892.

Walters, R. F., 1969. Contouring by machine: a user's guide. *The American Association of Petroleum Geologists Bulletin*, 53(11):2324–2340.

Warton, D. I., Duursma, R. A., Falster, D. S., and Taskinen, S., 2012. smatr 3–an R package for estimation and inference about allometric lines. *Methods in Ecology and Evolution*, 3:257–259.

Warton, D. I., Wright, I. J., Falster, D. S., and Westoby, M., 2006. Bivariate line-fitting methods for allometry. *Biological Reviews*, 81(2):259–291.

Wasserstein, R. L. and Lazar, N. A., 2016. The ASA's statement on p-values: Context, process, and purpose. *The American Statistician*, 70(2):129–133.

Weekley, R. A., Goodrich, R. K., and Cornman, L. B., 2010. An algorithm for classification and outlier detection of time-series data. *Journal of Atmospheric and Oceanic Technology*, 27(1):94–107.

Welch, P. D., 1967. The use of fast fourier transform for the estimation of power spectra: A method based on time averaging over short, modified periodogram. *IEEE Transactions on Audio Electroacoustics*, AU-15:70–73.

Wessel, P., Smith, W. H. F., Scharroo, R., Luis, J. F., and Wobbe, F., 2013. Generic mapping tools: improved version released. *Transactions, American Geophysical Union*, 94:409–410.

Wickham, H., 2007. Reshaping data with the reshape package. *Journal of Statistical Software*, 21(12):1–20.

Wickham, H., 2009. *ggplot2: elegant graphics for data analysis*. Springer, New York, USA.

Wickham, H., 2011. The split-apply-combine strategy for data analysis. *Journal of Statistical Software*, 40(1):1–29.

Wickham, H., 2014. *Advanced R*. The R Series. Chapman and Hall/CRC.

Wilson, J. T., 1963. Evidence from islands on the spreading of ocean floors. *Nature*, 197(4867):536–538.

Wilson, T., 1962. Cabot Fault, an Appalachian equivalent of the San Andreas and Great Glen Faults and some implications for continental displacement. *Nature*, 195(4837):135–138.

Wood, S. N., 2001. mgcv: GAMs and generalized ridge regression for R. *R News*, 1(2):20–25.

Wright, D. G., Pawlowicz, R., McDougall, T. J., Feistel, R., and Marion, G. M., 2011. Absolute salinity, "density salinity" and the reference-composition salinity scale: present and future use in the seawater standard TEOS-10. *Ocean Science*, 7:1–26.

Wu, A., Hsieh, W. W., and Tang, B., 2006. Neural network forecasts of the tropical Pacific sea surface temperatures. *Neural Networks*, 19(2):145–154.

Wunsch, C., 2015. *Modern Observational Physical Oceanography: Understanding the Global Ocean*. Princeton University Press, Princeton NJ.

Wyrtki, K., 1962. The oxygen minima in relation to ocean circulation. *Deep-Sea Research and Oceanographic Abstracts*, 9(1–2):11–23.

Xu, F. and Perrie, W., 2012. Extreme waves and wave run-up in halifax harbour under climate change scenarios. *Atmosphere-Ocean*, 50(4):407–420.

Zeileis, A., Hornik, K., and Murrell, P., 2009. Escaping RGBland: Selecting colors for statistical graphics. *Computational Statistics and Data Analysis*, 53(9):3259–3270.

Zika, J. D. and McDougall, T. J., 2008. Vertical and lateral mixing processes deduced from the Mediterranean Water signature in the North Atlantic. *Journal of Physical Oceanography*, 38(1):164–176.

Index

© Springer Science+Business Media, LLC, part of Springer Nature 2018
D. E. Kelley, *Oceanographic Analysis with R*,
https://doi.org/10.1007/978-1-4939-8844-0

Printed in the United States
By Bookmasters